科技英语丛书

普通高等学校"十三五"省级规划教材

化学专业英语基础教程

An Elementary Course of Chemistry English

第2版

主　编　赵建军
副主编　张现峰

中国科学技术大学出版社

内 容 简 介

本书主要面向初次接触化学专业英语的学生以及化学、化工英语的初学者,共包括三部分内容:第一部分为命名知识,讲述常见无机化合物和有机化合物基本构词知识及英文命名方法,采用中文形式,使内容更易于理解记忆,又以英文文章的形式补充讲述了构词法的一些基本常识;第二部分为化学基础知识,着重讲述了无机化学、有机化学、分析化学、物理化学等学科的发展历史及研究内容,突出科普性,着力培养学习化学专业英语人员的阅读兴趣;第三部分为文献导读,主要介绍文献的基本结构、写作要领及科技英语一般性表达方法,为以后撰写科技论文打下基础。

本书基础性强,注重实用,可供化学、应用化学、化工、轻工等专业的大学本科及专科学生使用。

图书在版编目(CIP)数据

化学专业英语基础教程/赵建军主编.—2版.—合肥:中国科学技术大学出版社,2018.2(2024.7重印)
ISBN 978-7-312-04387-1

Ⅰ.化… Ⅱ.赵… Ⅲ.化学—英语—高等学校—教材 Ⅳ.H06

中国版本图书馆 CIP 数据核字(2018)第 029610 号

出版	中国科学技术大学出版社 安徽省合肥市金寨路96号,230026 http://press.ustc.edu.cn https://zgkxjsdxcbs.tmall.com
印刷	合肥市宏基印刷有限公司
发行	中国科学技术大学出版社
经销	全国新华书店
开本	710 mm×1000 mm 1/16
印张	10.5
字数	253 千
版次	2011年9月第1版 2018年2月第2版
印次	2024年7月第8次印刷
定价	28.00 元

Preface
前　　言

随着全球化进程的不断发展,国际、校际间的教科研交流愈来愈频繁。专业英语的地位和作用已逐渐为人们所接受,并愈来愈受到人们的重视。当前,国内部分高校在化学教学中已使用双语进行教学,英文已逐渐作为与中文共同存在的第二"专业语言",而不是现在的专业"外语"。在这种环境之下,我们编写了这本《化学专业英语基础教程》。

本书是为化学、应用化学、化工、轻工等专业的大学本科及专科学生编写的一本专业英语教程。娴熟地掌握化学专业英语知识是了解与化学相关的科学研究内容的基本功之一。同时,化学专业英语也有着自己的专业表达特色,掌握丰富的化学英文词汇构成知识,能够锻炼准确迅速阅读文献的能力,也能够提高化学及相关专业英语论文的表达技巧。在此指导思想之下,我们根据自己的教学经验以及当前高校教学改革不断深入的实际情况,在原来讲义的基础之上,数易其稿而编成本书。

本书主要包括三部分内容:第一部分,命名知识。讲述常见无机化合物及有机化合物基本构词知识及英文命名方法,采用中文形式,使内容易于理解记忆,又以英文文章的形式补充讲述了构词法的一些基本知识。第二部分,化学基础知识。由于学习本课程的同学不但有一定的公共外语基础,而且有一定的化学基础知识,为了增加同学们阅读化学专业英语的兴趣,我们重点讲述无机化学、有机化学、分析化学和物理化学等基本知识。在实际教学过程中,教师可根据课时安排等实际情况有选择性地进行讲解。第三部分,文献导读。这部分内容主要是为了满足学生将来的实际需要而设置的,讲述一般专业英语论文的写作及与科研之间的内在联系,文献的基本结构、写作要领及科技英语的一般性表达方法,同时选择数篇国外著名期刊上的专业英语文献作为范例。第三部分旨在强化化学专业英语学习的实用性,培养学生积极阅读专业期刊的习惯,了解当前国际化学及相关科学的研究前沿以及热点。

本书第一部分、第三部分内容由赵建军老师编写，其中第三部分为了展示不同学术期刊的编排风格，基本保留了原刊物的论文风格。第二部分内容主要由张现峰老师编写。最后，赵建军老师对整本书的内容进行了编排。在本书编写过程中得到了浙江师范大学罗孟飞教授，中国矿业大学魏贤勇教授、宗志敏教授，金华职业技术学院宋宇鹏老师，南京师范大学化学与环境科学学院李利老师以及美国哈佛大学 George M. Whitesides 教授等的支持，在此特表示感谢。同时，部分内容参考了互联网上的一些资料，在此对这些资料的作者表示深深的敬意和万分的感谢！

由于编者水平的限制，书中难免存在一些不当之处，恳切希望广大读者予以批评指正。

编 者

2017 年 8 月

Contents
目 录

Preface 前言 ·· i

Section Ⅰ Nomenclature
第一部分 命 名 知 识

Nomenclature of Inorganic Compounds（1） 无机化合物的命名(1) ········ 3
Nomenclature of Organic Compounds（1） 有机化合物的命名(1) ········· 7
Nomenclature of Inorganic Compounds（2） 无机化合物的命名(2) ··············· 14
Nomenclature of Organic Compounds（2） 有机化合物的命名(2) ········· 18

Section Ⅱ Chemical Knowledge
第二部分 化学基础知识

Inorganic Chemistry 无机化学 ·· 25
Organic Chemistry 有机化学 ··· 31
Analytical Chemistry 分析化学 ·· 35
Physical Chemistry 物理化学 ··· 42
Macromolecular Chemistry 大分子化学 ··· 50
Material Chemistry 材料化学 ··· 53
Computational Chemistry 计算化学 ·· 57
Chemical Laboratory 化学实验室 ·· 66

Section Ⅲ Literature Introduction
第三部分 文 献 导 读

Whitesides' Group: Writing a Paper ·· 75
Identification of CuO species in high surface area CuO—CeO_2 catalysts
  and their catalytic activities for CO oxidation ··· 83
An evidence for the strong association of N-methyl-2-pyrrolidinone
  with some organic species in three Chinese bituminous coals ·········· 103

Acrylonitrile synthesis from acetonitrile and methanol over
  MgMn/ZrO$_2$ catalysts ·· 117

Appendices
附　　录

Translation of Literature Introduction　文献导读部分参考译文 ·········· 135
Names of the Chemical Elements　化学元素名称·················· 154
Names of the Frequently-used Chemical Experiment Instruments
  常用化学实验仪器名称 ·· 156
References　参考文献 ·· 157

Section I Nomenclature

第一部分 命名知识

Nomenclature of Inorganic Compounds (1)

无机化合物的命名(1)

在具体学习化学专业英语命名之前,大家首先要了解化学元素的英文名称(请参阅附录部分)。在口语表达中,分子式阅读顺序与中文有所区别:在英语中化学式的读法一般是从左往右,也就是按照其书写的顺序读取。

无机化合物包括酸、碱、盐以及其他化合物、水合物等,相应地,其命名也包括这几个方面。

1. 阴、阳离子的命名

(1) 阳离子的命名

阳离子包括单原子阳离子(如 Na^+,Ca^{2+} 等)和多原子阳离子(如 NH_4^+)。

对于单原子阳离子(monatomic cations),其命名方式为:元素 + ion。如果某元素能形成一种以上的阳离子,则使用斯托克数字(Stock number)来表示其所带电荷(只形成一种阳离子的不必用)。例如:

Na^+:sodium ion Ca^{2+}:calcium ion Al^{3+}:aluminum ion

Fe^+:iron(Ⅰ) ion Fe^{2+}:iron(Ⅱ) ion Fe^{3+}:iron(Ⅲ) ion

对于多原子阳离子(polyatomic cations),其命名方式为:原子团名称 + ion。例如:

NH_4^+:ammonium ion

(2) 阴离子的命名

阴离子包括单原子阴离子、含氧酸根阴离子以及含氢阴离子等。

对于单原子阴离子(monatomic anions),其命名方式为:元素名称词干 + -ide(后缀) + ion。例如:

F^-:fluoride ion Br^-:bromide ion I^-:iodide ion

氰根(CN^-)和氢氧根(OH^-)在命名时视同单原子阴离子。例如:

CN^-:cyanide ion OH^-:hydroxide ion

对于含氧酸根阴离子(oxyanions),其命名方式为:非氧元素词干 + -ate(后缀) + ion,我们可以将其译为"酸根离子"。例如:

CO_3^{2-}:carbonate ion　　SO_4^{2-}:sulfate ion

NO_3^-:nitrate ion　　ClO_3^-:chlorate ion

如果某元素能形成一种以上的含氧酸根阴离子，则其命名按以下规则进行：

① 高(过)酸根离子：per-(前缀) + 非氧元素词干 + -ate(后缀) + ion。例如：

ClO_4^-(高氯酸根离子)：perchlorate ion

② 酸根离子：非氧元素词干 + -ate(后缀) + ion。

③ 亚某酸根离子：非氧元素词干 + -ite(后缀) + ion。例如：

SO_3^{2-}(亚硫酸根离子)：sulfite ion　　NO_2^-(亚硝酸根离子)：nitrite ion
ClO_2^-(亚氯酸根离子)：chlorite ion

④ 次某酸根离子：hypo-(前缀) + 非氧元素词干 + -ite(后缀) + ion。例如：

ClO^-(次氯酸根离子)：hypochlorite ion

⑤ 偏某酸根离子：meta-(前缀) + 非氧元素词干 + -ate(后缀) + ion。例如：

PO_3^-(偏磷酸根)：metaphosphate ion

⑥ 焦某酸根离子：pyro-(前缀) + 非氧元素词干 + -ate(后缀) + ion。

⑦ 含氢阴离子(anions containing hydrogen)的命名方式为：hydrogen + 离子名称。例如：

HCO_3^-(碳酸氢根)：hydrogen carbonate ion

2. 酸的命名

酸的命名分为两类，即含氧酸的命名和非含氧酸的命名。

对于含氧酸的命名，其基本构词法为：酸根离子中非氧元素词干 + -ic(后缀) + acid。例如：

H_2SO_4(硫酸)：sulfuric acid　　H_3PO_4(磷酸)：phosphoric acid

若某元素能形成一种以上的含氧酸，则其命名按以下规则进行：

① 高(过)某酸：per-(前缀) + 酸根离子中非氧元素词干 + -ic(后缀) + acid。例如：

$HClO_4$(高氯酸)：perchloric acid

② 某酸：酸根离子中非氧元素词干 + -ic(后缀) + acid。

③ 亚某酸：酸根离子中非氧元素词干 + -ous(后缀) + acid。例如：

H_2SO_3(亚硫酸)：sulfurous acid　　$HClO_2$(亚氯酸)：chlorous acid

④ 次某酸：hypo-(前缀) + 酸根离子中非氧元素词干 + -ous(后缀) + acid。例如：

$HClO$(次氯酸)：hypochlorous acid

⑤ 偏某酸:meta-(前缀) + 酸根离子中非氧元素词干 + -ic(后缀) + acid。例如：

HPO_3（偏磷酸）: metaphosphoric acid

⑥ 焦某酸:pyro-(前缀) + 酸根离子中非氧元素词干 + (-ic) + acid。

从含氧酸阴离子以及含氧酸的命名我们可以看出：a. 对于可以显示不同价态的元素，其含氧酸和含氧酸根阴离子可以用不同的前后缀组合加以表示；b. 对于价态相同的元素，其含氧酸及含氧酸根阴离子具有相同的前缀、不同的后缀。

另外，其他的前缀还有：ortho-(正), thio-(硫代)。

对于非含氧酸的命名，其基本构词法为：hydro-(前缀) + 阴离子名称的词干 + -ic(后缀) + acid。例如：

HF(氢氟酸): hydrofluoric acid　　HCl(盐酸): hydrochloric acid

HBr(氢溴酸): hydrobromic acid　　H_2S(硫化氢): hydrosulfuric acid

HCN(氰化氢): hydrocyanic acid

3. 碱的命名

碱的命名原则为：元素名称 + hydroxide。若某元素能形成一种以上的阳离子，则使用斯托克数字来表示其所带电荷(只形成一种阳离子的不必用)。例如：

NaOH: sodium hydroxide　　$Fe(OH)_2$: iron(Ⅱ) hydroxide

4. 盐的命名

正盐的基本命名方法为：不带"ion"的阳离子名称 + 不带"ion"的阴离子名称。若某元素能形成一种以上的阳离子，则该阳离子的电荷数用斯托克数字来表示(只形成一种阳离子的元素不必用)。例如：

CuCl: copper(Ⅰ) chloride　　$CuCl_2$: copper(Ⅱ) chloride

$CuSO_4$: copper(Ⅱ) sulfate　　$KClO_4$: potassium perchlorate

酸(碱)式盐的基本命名方法为：不带"ion"的阳离子名称 + hydrogen(hydroxide) + 非氧元素词干 + -ate(后缀)，氢原子(氢氧根)的个数用前缀表示。例如：

$NaHCO_3$: sodium hydrogencarbonate

NaH_2PO_4: sodium dihydrogenphosphate

复盐的命名方法为：阳离子名称 + 阴离子名称 + 非氧元素词干 + -ate(后缀)。例如：

$KNaCO_3$: potassuim sodium carbonate

5. 分子化合物的命名

分子化合物的命名方法为:正价元素名称 + 负价元素名称词干 + -ide(后缀),分子中各元素原子的个数用数字前缀来表示。例如:

CaO: calcium oxide　　CO_2: carbon dioxide
CO: carbon monoxide　　P_2O_5: diphosphorus pentoxide
SF_6: sulfur hexafluoride

6. 水合物的命名

水合物的命名方法为:非水化合物名称 + 数字(结晶水个数)前缀 + hydrate。

常用数字前缀有:

1. mono-　　　　　2. bi-　　　　　3. tri-
4. tetra-　　　　　5. penta-　　　6. hex(a)-/sex(a)-
7. hept(a)-/sept(a)-　8. oct(a)-　　　9. non(a)-
10. dec(a)-

例如:

$CuSO_4 \cdot 5H_2O$: copper(Ⅱ) sulfate pentahydrate
$AlCl_3 \cdot 6H_2O$: aluminum chloride hexahydrate(或 aluminum chloride 6-water)

Nomenclature of Organic Compounds（1）

有机化合物的命名(1)

1. 烃类命名法

烃类化合物包括脂肪烃和芳香烃，其命名方式从这两个方面分别讲述。

(1) 脂肪烃的命名

脂肪烃命名，由表示碳原子个数的数字头加表示烃类的后缀构成。例如：十三烷烃，由表示 13 的数字头 "trideca-" 加表示烷烃的后缀 "-ane" 组合而成，即 tridecane。因此，表示碳原子个数的数字前缀大家一定要牢牢掌握。

① 数字前缀

当碳原子总数≤10 时：

  甲:meth- 乙:eth- 丙:prop- 丁:buta- 戊:penta-
  己:hexa- 庚:hepta- 辛:octa- 壬:nona- 癸:deca-

另有：

  半,1/2:hemi-, semi- 单,一:mono-, uni-
  3/2:sesqui- 双,两:di-, bi-, bis-

当碳原子总数＞10 时：数字由表示个位数字的 1,2,3,4 等加上表示十位数字的 10,20,30,40 等构成，例如：13 是由表示个位数字的 3 和表示十位数字的 10 构成的。

在这种构成方式中，个位数字表示为：

  hen(i)- do- tri(a)- tetra- penta-
  hexa- hepta- octa- nona-

十位数字表示为：

  deca- eicosa- triaconta- tetraconta- pentaconta-
  hexaconta- heptaconta- octaconta- enneaconta-

例如：

  hendeca-(有时也用 undeca-) dodeca- trideca-
  tetradeca- pentadeca- hexadeca- heptadeca-
  octadeca- nonadeca- eicosa- heneicosa-

　　　　　docosa-　　　　tricosa-　　　　tetracosa-　　　pentacosa-
　　　　　hexacosa-　　　heptacosa-　　　octacosa-　　　 nonacosa-
　　　　　triaconta-　　　hentriaconta-

② 烷烃的命名

烷烃的命名由表示碳原子个数的数字加后缀"-ane"表示，即：数字头加表示烷烃的后缀"-ane"，其中以 a 结尾的数字头(4 以上)直接加"-ne"。例如：

　　甲烷：methane　　　　　庚烷：heptane　　　　　癸烷：decane
　　十三烷：tridecane　　　　十四烷：tetradecane　　　十五烷：pentadecane
　　二十烷：(e)icosane　　　 二十一烷：heneicosane　　二十二烷：docosane
　　三十烷：triacontane　　　四十烷：tetracontane　　　五十烷：pentacontane
　　六十烷：hexacontane　　 七十烷：heptacontane　　　八十烷：octacontane
　　九十烷：nonacontane　　 一百烷：empirecontane

③ 烯烃的命名

其命名方式和烷烃相似：数字头加表示烯烃的后缀"-ene"，若数字头以 a 结尾，则将数字头结尾的"a"去掉，再加"-ene"。例如：

　　　　　　　　乙烯：ethene　　丁烯：butene

对于多烯的命名，用后缀"-diene"表示二烯类，用后缀"-triene"表示三烯类，下面以此类推。前面再冠以数字头。例如：

　　　　　　　丁二烯：butadiene　丁三烯：butatriene

④ 炔烃的命名

数字头加表示炔烃的后缀"-yne"(有时用-ine)。二炔类：数字头 + -diyne，以 a 结尾的数字头去"a"加"-yne"。例如：

　　　　乙炔：ethyne　丁炔：butine　己二炔：hexadiyne 或 hexadine

⑤ 脂环烃的命名

cyclo-(前缀) + 烃类名称。例如：

　　　　　　环己烷：cyclohexane　环己二烯：cyclohexadiene

⑥ 含有烃基的支链烃类的命名

烃基的命名，烷基的英文名称是将烷烃的词尾"ane"改为"yl"，如：甲基 methyl,乙基 ethyl 等。烯基及炔基的英文名称则是相应烃末尾字母"e"去掉再加"yl"。如乙烯基为 ethenyl,乙炔基为 ethynyl,丙烯基为 propenyl(烯丙基：allyl)。

主链选取以及碳原子编号原则和我们在有机化学中所学习内容一样。但是应当注意以下几点：a. 在英文书写时，侧链取代基的书写顺序按照取代基字头的英文顺序书写；b. 当烯烃(-ene)或炔烃(-yne)前还有相同的不饱和键时，用"di-"或"tri-"等字头表示，如：二烯烃，命名为"-diene"；c. 若烃类中同时含有双键和三键，用"-en"或"-yne"形式作为结尾。例如：

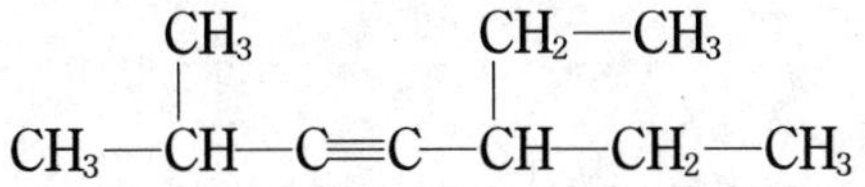

$$\text{CH}_3\text{—CH}\underset{|}{\overset{\overset{\text{CH}_3}{|}}{\text{—}}}\text{C}\equiv\text{C—CH}\underset{}{\overset{\overset{\text{CH}_2\text{—CH}_3}{|}}{\text{—}}}\text{CH}_2\text{—CH}_3$$

2-甲基-5-乙基-3-庚炔：5-ethyl-2-methyl-3-heptyne

$$\text{CH}_3\text{—CH}=\text{CH—CH}=\text{CH—CH}_3$$

(2E,4E)-2,4-己二烯：(2E,4E)-2,4-hexadiene

$$\text{CH}\equiv\text{C—CH}_2\text{CH}_2\text{—CH}=\text{CH}_2$$

1-己烯-5-炔：hex-1-en-5-yne

(2) 芳香烃的命名

芳香烃是以苯环作为母体的烃类，一般以苯(benzene)作为母体，其他作为取代基。若以苯作为取代基，根据上面所述，应写为"phenyl-"。例如：

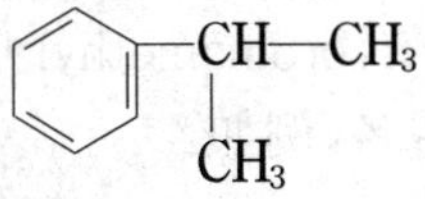

异丙基苯：isopropyl benzene

2. 其他有机物的命名

有机化合物又称官能团化合物，足见官能团在有机化合物中的重要性。在有机化合物命名时也是这样，首先选择主要的官能团。常见的官能团及其中英文名称如下：

—COOH：羧基，carboxy　　—COOR：烃氧甲酰基，R-oxycarbonyl
—COX：卤甲酰基，halocarbonyl　　—COF：氟甲酰基，fluroformyl
—COCl：氯甲酰基，chloroformyl　　—COBr：溴甲酰基，bromoformyl
—COI：碘甲酰基，iodoformyl　　—CONH$_2$：氨基甲酰基，carbamoyl
—CN：氰基，cyano　　—CHO：甲酰基(醛基)，formyl
—C=O：羰基(氧代)，oxo　　—OH：羟基，hydroxy
—NH$_2$：氨基，amino-　　—NO$_2$：硝基，nitro-
—NO：亚硝基，nitroso　　—OR：烃氧基，R-oxyp

在烷氧基中，低于5个碳时将烷基中英文词尾"yl"省略，例如：

甲氧基：methoxy　　苄氧基：benzyloxy

官能团的优先顺序在IUPAC中有详细规定，位置在前的官能团优先，可作为主要的官能团(称为相应的化合物，如：醇、醚等)，其余的作为取代基，在命名时，连同其所在碳原子编号一起放于字头。下面分别予以讲述。

(1) 卤化物的命名

卤化物既是由卤素作为取代基的有机物,在命名时,只需将卤元素名称名后面的"ine"去掉,改成"o"即可。如:fluorine→fluoro-,chlorine→chloro-,bromine→bromo-,iodine→iodo-。具体例如:

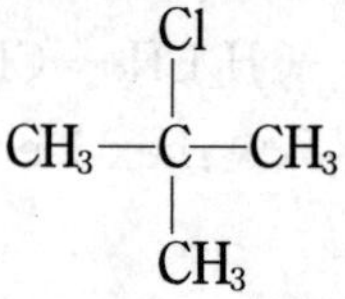

$$CH_3-\underset{\underset{CH_3}{|}}{\overset{\overset{Cl}{|}}{C}}-CH_3$$

2-氯-2-甲基丙烷:2-chloro-2-methyl propane

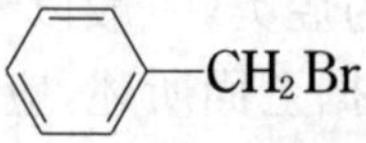

$$C_6H_5-CH_2Br$$

溴甲基苯:bromomethyl benzene

同时,大家要记住一些物质的俗名,例如:

氯仿:chloroform　溴仿:bromoform　氟里昂:freon(氟、氯代烷)

(2) 醇、酚的命名

烃类物质的英文名称是以字母"e"结尾的,醇、酚的命名即在相应烃的名称后,将"-e"去掉再加"-ol",称为相应的醇或酚。如果是二醇或三醇,则相应变为"-diol"或"-triol"(也有不去"-e"的)。例如:

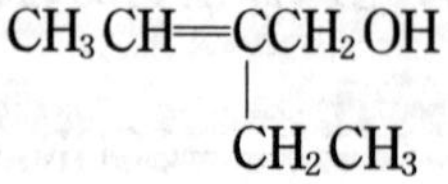

$$CH_3CH=CCH_2OH$$
$$\quad\quad\quad\quad |$$
$$\quad\quad\quad\quad CH_2CH_3$$

2-乙基-2-丁烯-1-醇:2-ethyl-2-buten-1-ol(2-ethyl-but-2-en-1-ol)

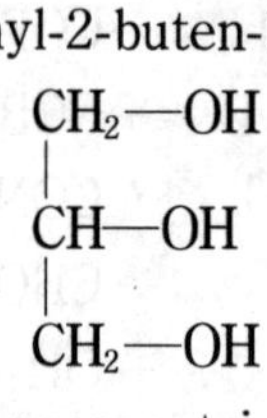

$$CH_2-OH$$
$$|$$
$$CH-OH$$
$$|$$
$$CH_2-OH$$

1,2,3-丙三醇:1,2,3-propanetriol(俗称甘油,glycerin)

$$HO-C_6H_4-OH$$

1,4-苯二酚:1,4-benzendiol

(3) 醚的命名

将较简单的烃基与氧原子一起作为取代基命名,也就是以烃氧基作为取代基加以命名。有些简单醚的命名由连接氧原子的两个烃基加"醚"(ether)字组成,即"某某醚",其中两个烃基顺序与取代基顺序相同,例如:

$$CH_3-CH_2-O-CH_3$$

甲乙醚：methoxy ethane(或 ethyl methyl ether)

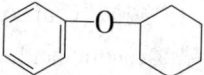

环己基苯基醚：cyclohexoxy benzene(或 cyclohexyl phenyl ether)

对于环醚，则以烃基为母体，并在母体前加前缀"epoxy-"，同时标出与氧原子相连的两个碳原子相应编号。例如：

$$\begin{array}{c} O \!-\! CH_2 \\ | \quad\; | \\ CH_2 \!-\! CH_2 \end{array}$$

1,3-环氧丙烷：1,3-epoxypropane

$$\begin{array}{c} CH_2 \!-\! CH_2 \\ | \qquad | \\ CH_2 \quad CH_2 \\ \diagdown \;\; \diagup \\ O \end{array}$$

1,4-环氧丁烷：1,4-epoxybutane[俗称四氢呋喃(THF)，tetrahydrofuran]

(4) 醛的命名

醛的命名是将同样碳数烃的名称后去"-e"加"-al"。例如：

HCHO

甲醛：methanal

$$\begin{array}{c} CHO \\ | \\ CH_3CHCH_3 \end{array}$$

2-甲基丙醛：2-methyl propanal

(5) 酮的命名

酮的命名是将同样碳数烃的名称后去"-e"加"-one"。例如：

$$\begin{array}{c} CH_3CCH_2CH_2CH_3 \\ \| \\ O \end{array}$$

2-戊酮：2-pentanone(或甲基丙基酮，methyl propyl ketone)

$$\begin{array}{c} CH_3CCH_3 \\ \| \\ O \end{array}$$

丙酮：propanone(或二甲基酮，dimethyl ketone)

(6) 羧酸的命名

将相同碳原子个数的烃去"-e"加"-oic acid"。编号从—COOH 上的碳原子开始(适用于链状的一元或二元酸)。若为二元羧酸用"dioic acid"，三元羧酸用

"trioic acid"加在烃名称的后面,同时保留烃字尾"e"。若羧酸基直接连在脂环上,则命名时在脂环烃的后面加上"羧酸"(carboxylic acid)。

一些酸还保留有俗名。如:醋酸,acetic acid;苯甲酸,benzoic acid 等。例如:

$$CH_3CH_2CH_2COOH$$

丁酸:butanoic acid

$$CH_3CH\!\!=\!\!CHCH_2COOH$$

3-戊烯酸:3-pentenoic acid

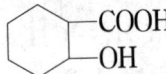

2-羟基-1-环己烷羧酸:2-hydroxy-1-cyclohexane carboxylic acid

(7) 酰卤的命名

在中文命名时,酰卤的名称是酰基加上卤素。如:丙酰氯、甲酰氯等。在英文命名时,只需将相应羧酸名称后的"-ic acid"换成"yl",再加上 bromide 或 chloride。二酰用相应的烷烃加"dioyl"。例如:

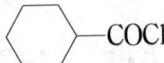

氯代环己烷甲酰:cyclohexane carboxyl chloride

(8) 酯类的命名

酯类物质是由酸和醇反应而形成的,在命名时,将酸部分去掉"-ic acid"加"-ate",前面再加上烃基,并隔开。例如:

$$CH_3COOCH_3$$

乙酸甲酯:methyl acetate(或 methyl ethanate)

$$CH_3CH_2COOC_2H_5$$

丙酸乙酯:ethyl propanoate

(9) 酰胺的命名

此类物质的命名,是将相应羧酸名称后的"-oic acid"改为"-amide",或者将相应羧酸名称后的"carboxylic acid"改成"carboxamide"。例如:

$$CH_3(CH_2)_3CONH_2$$

戊酰胺:pentanamide(或 butane carboxamide)

(10) 胺类的命名

胺的命名是烃基后面加"胺"字,如甲胺、苯胺等。其英文命名即是用"amine"代替烃中词尾"e"。若为二胺或三胺则分别用"diamine"或"triamine"命名。例如:

$$\begin{array}{c} \text{CH}_3 \\ | \\ \text{CH}_3\text{—N—(CH}_2)_3\text{—CH}_3 \end{array}$$

N,N-二甲基丁胺:N,N-dimethyl butyl amine

(11) 硝基化合物的命名

硝基化合物在命名时,将硝基看成取代基,烃看成母体,如硝基苯。英文名称硝基是"nitro"。例如:

$$\text{CH}_3\text{NO}_2$$

硝基甲烷:nitromethyl

Nomenclature of Inorganic Compounds (2)

无机化合物的命名(2)

1. Naming elements

Each element has a symbol that is used by chemists around the world as a kind of shorthand. Symbols for elements may have one letter or two.

Wherever possible, the symbol is the first letter of the common name or the Latin name of the element. For example, the symbol for hydrogen is H; for carbon, C; for uranium, U. The symbol for potassium is K, after kalium, the Latin name of one of their compounds, the elements themselves having been discovered only in relatively recent times.

Symbols for elements may be found in Names of the Chemical Elements.

2. Naming Acids

There are TWO types of acids:
1) Binary acids (acids that do not contain oxygen atom);
2) Oxo-acids (those containing oxygen atoms in their formula).

The names of binary acids start with hydro-, followed by the first syllable of the anion's name, and end with -ic:

HF: hydrofluoric acid   HCl: hydrochloric acid   HBr: hydrobromic acid
HI: hydriodic acid H_2S: hydrosulfuric acid HCN: hydrocyanic acid

The names of oxyacids are derived from the name of oxy-anion the acids contain. For an anion whose name ends with -ate, the acid's name starts with the first syllable of the anion and ends with -ic. If the anion's name ends with -ite, the name of acid starts with the first syllable of the anion's name and ends with -ous.

Table 1   Names of Anions and Acids

Anions	Names of Anions	Acids	Names of Acids
NO_3^-	nitrate ion	HNO_3	nitric acid
NO_2^-	nitrite ion	HNO_2	nitrous acid
SO_4^{2-}	sulfate ion	H_2SO_4	sulfuric acid
SO_3^{2-}	sulfite ion	H_2SO_3	sulfurous acid
$C_2H_3O_2^-$	acetate ion	$HC_2H_3O_2$	acetic acid
ClO^-	hypochlorite	$HClO$	hypochlorous acid
ClO_2^-	chlorite	$HClO_2$	chlorous acid
ClO_3^-	chlorate	$HClO_3$	chloric acid
ClO_4^-	perchlorate	$HClO_4$	perchloric acid

Note: the ionizable hydrogen in oxo-acids is bonded to the oxygen in the molecule.

3. Naming Metal Oxides, Bases and Salts

A compound is a combination of positive and negative ions in the proper ratio to give a balanced charge and the name of the compound follows from names of the ions, for example, NaCl is sodium chloride; $Al(OH)_3$ is aluminium hydroxide; $FeBr_2$ is iron (Ⅱ) bromide or ferrous bromide; $Ca(OAc)_2$ is calcium acetate; $Cr_2(SO_4)_3$ is chromium (Ⅲ) sulphate or chromic sulphate, and so on. The name of the negative ion can be shown in Table 2. Table 3 gives some examples of naming metal compounds.

Table 2   Some Common Negative Ions

Symbol	Name	Symbol	Name
NO_3^-	nitrate	NO_2^-	nitrite
CO_3^{2-}	carbonate	SO_3^{2-}	sulphite
SO_4^{2-}	sulphate	PO_3^{3-}	phosphite
PO_4^{3-}	phosphate	AsO_3^{3-}	arsenite
HSO_4^-	hydrogen sulphate	HSO_3^-	hydrogen sulphite
HCO_3^-	hydrogen carbonate	ClO^-	hypo-chlorite

(Continued)

Symbol	Name	Symbol	Name
AsO_4^{3-}	arsenate	CN^-	cyanide
IO_3^-	iodate	I^-	iodide
ClO_3^-	chlorate	F^-	fluoride
CrO_4^-	chromate	Cl^-	chloride
$Cr_2O_7^{2-}$	dichromate	Br^-	bromide
ClO_4^-	perchlorate	S^{2-}	sulphide
MnO_4^-	permanganate	O^{2-}	oxide
OAc^-	acetate	H^-	hydride
$C_2O_4^{2-}$	oxalate	OH^-	hydroxide

Table 3 Names of Some Metal Oxides, Bases and Salts

Formula	Name	
FeO	iron(II) oxide	ferrous oxide
Fe_2O_3	iron(III) oxide	ferric oxide
$Sn(OH)_2$	tin(II) hydroxide	stannous hydroxide
$Sn(OH)_4$	tin(IV) hydroxide	stannic hydroxide
Hg_2SO_4	mercury(I) sulphate	mercurous sulphate
$HgSO_4$	mercury(II) sulphate	mercuric sulphate
NaClO	sodium hypochlorite	
$K_2Cr_2O_7$	potassium dichromate	
$Cu_3(AsO_4)_2$	copper(II) arsenate	cupric arsenate
$Cr(OAc)_3$	chromium(III) acetate	chromic acetate

Negative ions, or anions, may be monatomic or polyatomic. All monatomic anions have names ending with -ide. Two polyatomic anions which also have names ending with -ide are the hydroxide ion, OH^-, and the cyanide ion, CN^-.

Many polyatomic anions contain oxygen in addition to another element. The number of oxygen atoms in such oxyanions is denoted by the use of the suffixes -ite and -ate, meaning fewer and more oxygen atoms, respectively. In cases where it is necessary to denote more than two oxyanions of the same element, the prefixes hypo- and per-, meaning still fewer and more oxygen atoms, respectively, may be used, as shown in the following:

hypochlorite ClO$^-$	chlorite ClO$_2^-$
chlorate ClO$_3^-$	perchlorate ClO$_4^-$

A salt containing acidic hydrogen is termed an acid salt. A way of naming these salts is to call Na$_2$HPO$_4$ disodium hydrogen phosphate and NaH$_2$PO$_4$ sodium dihydrogen phosphate. Historically, the prefix bi- has been used in naming some acid salts; in industry, for example, NaHCO$_3$ is called sodium bicarbonate and Ca(HSO$_3$)$_2$ calcium bisulphite. Bi(OH)$_2$NO$_3$, a basic salt, would be called bismuth dihydroxynitrate. NaKSO$_4$, a mixed salt, would be called sodium potassium sulphate.

4. Naming Nonmetal Oxides

One still widely used method employs Greek prefixes for both the number of oxygen atoms and that of the other elements in the compound. The prefixes used are (1) mono-, sometimes reduced to mon-, (2) di-, (3) tri-, (4) tetra-, (5) penta-, (6) hexa-, (7) hepta-, (8) octa-, (9) nona- and (10) deca-. Generally speaking, the letter "a" is omitted from the prefix (from tetra on) when naming a nonmetal oxide and "mono-" is usually omitted from the name altogether.

The Stock system is also used with nonmetal oxides. In either system, the element other than oxygen is named first, the full name being used, followed by oxide. Table 4 shows some examples.

Table 4 Names of Some Nonmetal Oxides

Formula	Name	
CO	carbon(II) oxide	carbon monoxide
CO$_2$	carbon(IV) oxide	carbon dioxide
SO$_3$	sulphur(VI) oxide	sulphur trioxide
N$_2$O$_3$	nitrogen(III) oxide	dinitrogen trioxide
P$_2$O$_5$	phosphorus(V) oxide	diphosphorus pentoxide
Cl$_2$O$_7$	chlorine(VII) oxide	dichlorine heptoxide

Nomenclature of Organic Compounds (2)

有机化合物的命名(2)

A complete discussion of definitive rules of organic nomenclature would require more space than can be allotted in this text. We will survey some of the more common nomenclature rules, both IUPAC and trivial.

1. Naming Alkanes and Cycloalkanes

Alkanes and cycloalkanes are the family of saturated hydrocarbons, that is, molecules containing carbon and hydrogen connected by single bond only. These molecules can be in continuous chains or in rings. The names of alkanes and cycloalkanes are the root names of organic compounds. Beginning with the five-carbon alkane, the number of carbons in the chain is indicated by the Greek or Latin prefix. Rings are designated by the prefix "cyclo". (In the geometrical symbols for rings, each apex represents a carbon with the number of hydrogens required to fill its valence.) The names for continuous chains alkanes are listed in Table 1.

Table 1 Names of Continuous-chain Alkanes

Formula	Name	Formula	Name
CH_4	methane	$C_{11}H_{24}$	undecane
C_2H_6	ethane	$C_{12}H_{26}$	dodecane
C_3H_8	propane	$C_{13}H_{28}$	tridecane
C_4H_{10}	butane	$C_{14}H_{30}$	tetradecane
C_5H_{12}	pentane	$C_{15}H_{32}$	pentadecane
C_6H_{14}	hexane	$C_{16}H_{34}$	hexadecane
C_7H_{16}	heptane	$C_{17}H_{36}$	heptadecane
C_8H_{18}	octane	$C_{18}H_{38}$	octadecane
C_9H_{20}	nonane	$C_{19}H_{40}$	nonadecane
$C_{10}H_{22}$	decane	$C_{20}H_{42}$	eicosane

2. Naming Alkenes and Alkynes

Alkenes and alkynes are hydrocarbons which respectively have carbon-carbon double bond and carbon-carbon triple bond functional groups. Unbranched hydrocarbons having one double bond are named in the IUPAC system by replacing the ending -ane of the alkane name with -ene. If there are two or more double bonds, the ending is -adiene, -atriene, etc.

Unbranched hydrocarbons having one triple bond are named by replacing the ending -ane in alkane with -yne. If there are two or more triple bonds, the ending is -adiyne, -atriyne, etc.

3. Naming Alcohols

All alcohols contain one —OH group at least. IUPAC names are taken from the name of the alkane with changing the final -e to -ol. In the case of polyols, the prefix such as di-, tri- etc., is placed just before -ol, with the position numbers placed at the front of the name, if possible, you can write the structural formula by yourself, such as, 1,4-cyclohexandiol. Here is an example:

$$H_3C-\overset{\overset{\displaystyle OH}{|}}{CH}-CH=CH_2$$

but-3-en-2-ol

4. Naming Carboxylic Acids

The Carboxylic Acid Family is a family of organic compounds with the functional group being the carboxyl group, —COOH. This group is attached to one of the carbons in the rest of the molecule. It is actually a carbonyl group, C=O, bonded to a hydroxyl group, OH. Carboxylic acids are named systematically from their corresponding alkanes, alkenes or alkynes by changing the ending -ane, -ene or -yne to -oic acid.

Example: $CH_3CH=CHCH_2COOH$: 3-pentenoic acid.

5. Naming Esters

Esters can be produced by an equilibrium reaction between alcohol and carboxylic acid. The ester is named according to the alcohol alkyl group and then the name of the acid while changing the "-ic acid" to "-ate." For example: CH_3COOCH_3: methyl acetate(or methyl ethanate).

6. Naming Acid Anhydrides

A compound that can be dissolved in water forming an acid solution is called an acid anhydride (acid without water).

Symmetrical acid anhydrides are named by replacing acid in the name of the corresponding acid with anhydride, such as acetic anhydride.

Unsymmetrical acid anhydrides are named by first naming each component carboxylic acid alphabetically arranged (without the word acid) followed by spaces and then the word anhydride, for example: acetic formic anhydride.

7. Naming Aldehydes

The systematic names for aldehydes are given by adding suffix -al to the name of the parent alkane. The parent chain must contain the CHO group, and the —CHO carbon is always numbered as carbon 1. For example: HCHO: methanal.

The common names of aldehydes are derived from the names of the corresponding carboxylic acids. For example: CH_3COOH: acetic acid; CH_3CHO: acetaldehyde.

For more complex aldehydes in which the CHO group is attached to a ring, the suffix -carbaldehyde is used, also.

8. Naming Ethers

Ethers are compounds with two alkyl groups bonded to one oxygen atom. The general structure for ethers is R-O-R', where R and R' are hydrocarbon groups. Symmetrical ethers are ethers where the alkyl groups R and R' are the same, and asymmetrical ethers possess the difference between R and R'.

Simple ethers can be named in the way that the alkyl groups alphabetically are followed by the word "ether". Another way of naming ethers would be by adding -oxy- to the prefix for the smaller hydrocarbon group and joining it to the alkane name of the larger hydrocarbon group. For example, CH_3—O—CH_2—CH_3 would be called using this common name approach as ethyl methyl ether and another way to name it would be methoxyethane. If just two alkyl groups are identical in its molecular formula, use the prefix di-, tri-, tetra-, etc., to signify these branches. For example: diethyl ether.

However, if more complex ethers that have branching, use this common name approach, it would be considerably more difficult to do that. The IUPAC have come up with some rules that allow the naming of complex ethers. The rules are similar to those used in naming alcohols except that the O—R group is named as any other branched group. Using the rules for alkanes, alkenes, or alkynes with the alkoxy groups identified on the longest continuous chain. For example:

$$CH_3CH_2—O—CH_2CH_3$$
ethoxyethane

9. Naming Ketones

In the systematic names for ketones, the -e of the parent alkane name is dropped and -one is added. A prefix number is used if necessary. Here is an example:

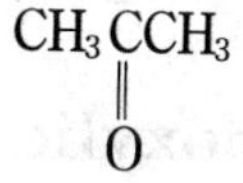

$$CH_3CCH_3$$
$$\parallel$$
$$O$$

dimethyl ketone (or propanone)

In a complex structure, a ketone group may be named in IUPAC system with the prefix oxo-. (The prefix keto- is also sometimes encountered.) For example:

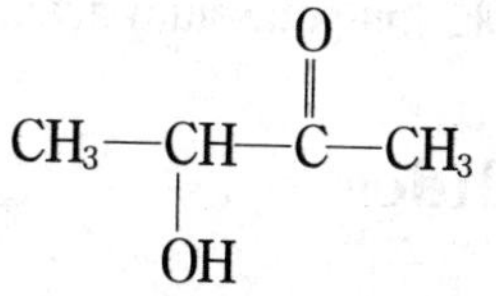

$$CH_3—CH—C—CH_3$$
with OH on the CH and =O on the C

3-hydroxybutan-2-one

10. Naming Amines

A primary amine contains the group —NH_2 attached to a hydrocarbon chain or ring. One can think of amines in general as being derived from ammonia, NH_3. In a primary amine, one of the hydrogens in an ammonia molecule has been replaced by a hydrocarbon group. In a secondary amine, two of the hydrogen atoms have been replaced by hydrocarbon groups. In a tertiary amine, all three hydrogens have been replaced.

Amines are named in two principal ways: with -amine as the suffix and with amino- as a prefix. Names for some ethers and amines can be found in Table 2.

Table 2 Names for Some Amines

Common	IUPAC
methylamine	aminomethane
diethyl amine	2-amino-butane
ethyl-methyl amine	dimethyl aminoethane

11. Naming Amides

Amides contain the group —$CONH_2$ where the —OH of an acid is replaced by —NH_2.

In both the IUPAC and trivial systems, an amide is named by dropping the ending -ic or -oic of the corresponding acid name and adding -amide, such as hexanamide (IUPAC) and acetamide (trivial).

12. Naming Salts of Carboxylic Acids

When carboxylic acids form salts, the hydrogen in the —COOH group is lost and replaced by a metal. So, carboxylic acid salts are named in both the common and IUPAC systems by replacing the -ic ending of the acid name with -ate. For example, $CH_3CO—OK^+$ is potassium acetate or potassium methanoate.

13. Naming Acid Halides

Acid halides, RCOX, are named by changing the suffix of the carboxylic from -ic acid to -yl, and then adding the corresponding halide name in the following. For example, acetyl chloride.

Section Ⅱ　Chemical Knowledge

第二部分　化学基础知识

Inorganic Chemistry

无 机 化 学

Inorganic chemistry is the branch of chemistry concerned with the constitute, property, structure and reaction of inorganic compounds, and is the most ancient branch academics in the chemistry. The inorganic matters almost include all chemical elements and their compounds, however, most of the carbon compounds are excluded.

In the past, people think inorganic matters were abiotic matters, such as the rock, soil, mineral, water, etc.. Organic material comes from zoetic animal and plant, like protein, grease, starch, cellulose, urea, etc.. In 1828, Reich chemist Verel prepared urea from the inorganic matter ammonium cyanate, and got rid of superstition of organic matter only produced by vitality, showing the signs that both inorganic and organic matters are integrated by the chemistry bond theory.

1. Inorganic Chemistry Development History

Hominid had the ability of recognizing inorganic matter in nature existence and make full use of them. Afterwards, people detected nature materials and found them can become new dissimilar materials by chance, hence taken into imitate, this was the beginning of ancient chemistry technics.

In the 6000 B. C., Chinese ancestors had known to burn clay to make pottery, and gradually developed to colourful pottery, white pottery, enamel pottery and china. Around 5000 B. C., mankind detected natural copper having tough and resilient ability, could be used to make tools which were not easily damaged. In the 2nd century B. C., Chinese people detected the reaction of iron with copper compounds forming copper metal, and this reaction became one of the subsequent methods of producing copper.

In the Shang dynasty in the 17th century B. C., people had known salt (sodium chloride) was a condiment. In the 5th century B. C., azure stone container had already been discovered. In the 7th century A. D., there were nitric(potassium nitrate), brimstony and charcoal which were made into powder in China. In the Ming Dynasty, Song Yingxing detailedly recorded handicraft technique in *Exploitation of the Works of Nature*: *Tian Gong Kai Wu* published in 1637, which included chinaware, copper, steel, salt, nitric, lime, red and yellow alum, and several other kinds of inorganic matter process. So we can see, before the establishment of chemistry science, the mankind have already controlled a great deal of the knowledge and technique of inorganic chemistry.

Ancient alchemy was the pioneer of chemistry science, people wanted to change cinnabar into gold, and refine the longevity pill. Knowledge concerning inorganic matter transformation was gained mainly from the experiment. Alchemist designed and made experiment tools, such as heat stove, reactor, distiller etc.. Although the alchemists' purpose were fantastic, operation method and the sensitive knowledge became the forerunner of chemistry science.

Because the initial study is on inorganic chemistry, so the establishment of modern inorganic chemistry marks the origin of modern chemistry. There are three chemists whose contributions are the most remarkable to the establishment of modern chemistry, namely British R. Boyle, French L. Lavoisier, British Dalton.

R. Boyle once carried on a lot of experiments on the chemistry, like phosphor, hydrogen preparation, the metals dissolved in acid and sulphur, hydrogen combustion. He elaborated differentiation of chemical elements and compounds from the experiment result, put forward that the chemical element is a kind of material that can't be separated into other materials. These new concepts and new standpoints, led chemistry research to the right way, made outstanding contribution to the establishment of modern chemistry.

L. Lavoisie adopt balance for importance tool of material variety research, and carried on many experiments, such as sulphur, phosphoric combustion, tin, mercury, metals heat in the air, etc.. He established exactitude concept that material combustion was an oxygenation, which overthrew phlogiston doctrine that had been widely accepted for about a hundred years at the time.

In 1803, Dalton put forward atom theory and declared the whole chemical element all constitute from particle that can't be parted, can't be demolished. After the atom theory establishment, chemistry science formally announce to establish.

In the 1830s, there were more than 60 kinds of chemical elements known by people. Russian chemist Mendeleyev researched on the properties of these chemical elements. In 1869, he put forward chemical element periodic law: the properties of chemical element present a periodic variety with the increment of the atomic weight of chemical element. This law provides a systematic classification of the nature of chemical elements. The periodic table of elements arranges all the chemical elements in a periodic tribe according to the periodic law, and researches on the periodic law did promote the development of inorganic chemistry.

The existing chemical elements that are already known are of 109 kinds in total, among which 94 elements can be found directly in the nature, and the rest 15 kinds are artificial. The signs used to represent chemical elements are mostly Latin abbreviation. Some Chinese names of chemical elements are familiar from immemorial years, like gold, aluminum, copper, iron, tin, sulphur, arsenic, phosphor, etc.. Some come from English transliteration, like sodium, manganese, uranium, helium, etc.. Also there are creations according to meaning, like hydrogen(light gas), bromine(smelly water), platinum(white gold), etc..

The periodic law is important for the development of chemistry. According to the periodic law, Mendeleyev once predicted that there were still lots of already existed chemical elements waiting to be discovered at that time. Periodic law still instructs systemic research of the chemical element and compound property, which become the stimulant to the foundation of modern material structure theories development.

A series of discovery at the end of 19th century contributed to the establishment of modern inorganic chemistry. In 1895, Roentgen discovered X ray; in 1896, Becquerel discovered radioactivity of the uranium; Thomson detected electrons in 1897; in 1898, Pierre Curie and his wife detected radioactivity of polonium and radium. At the beginning of 20th century, Rutherford and Bohr put forward an atom structure model constituted by atomic nucleus and electrons, which changed Dalton's atom theory that atom can't be separated any more.

In 1916, Kossel put forward the electricity bond theories, and Louis established the covalent bond theories, both of which successfully explained problems such as the valence of atoms and the structure of the compound etc.. In 1924, Maurice de Broglie thought the material particle, such as an electron, had wave-particle duality theory. In 1926, Erwin Schrödinger established wave equation about particle motion; and one year later, Heitler and London applied the quantum mechanics processing hydrogen molecule, proved electrons probability density convergence observably between two hydrogen nuclei, and put forward the modern standpoint of chemistry bond.

Henceforth, previous researches have done much to found the valence bond theory, molecular orbital theory and ligand field theory. These three basic theories were essential theories of modern inorganic chemistry.

2. The Research Subjects of Inorganic Chemistry

At the beginning of the establishment of inorganic chemistry, its subjects have covered four types, namely, fact, concept, laws and theory.

Material gained directly through observation by resorting to sense organs, is called fact. By analyzing, comparing, integrating and generalizing concrete ideas, we can get abstract conceptions, for example, chemical element, compound, chemical combination, decompound, oxidization, deoxidization, atoms, were concrete notions at first. Then laws can be formulated by combining corresponding concept with the same generalizing fact, for example, the dissimilar chemical elements combined into various compounds, summarizing their fixed amount relation, and then having gained the quality conservation, we achieved fixed ratio laws. Establishing new concepts to elucidate relevant laws, and testing the exactitude of new concepts through experiment, then theories can be formed. For example, atom theory could elucidate each law's relevant chemical element weight relation that were established already at that time.

These derivative relations of chemistry knowledge enunciate their internal contact. Laws generalize facts, theories explain and combine each law, and thus the entire chemistry subject is made up of a scientific system of knowledge. It is considered that modern chemistry is established after the foundation of Dalton atom theory, because this theory deals with temporal chemistry contents in a systematic way.

In order to gain systematical chemistry knowledge, we should carry on research according to scientific methods. Scientific methods are mainly divided into the following three steps.

(1) Collecting Facts

Methods of collection include observation and experiment. The experiment is observation under the manipulative condition. Chemistry research values much of experiments, because chemistry variety phenomena are all very complex in the nature, the essence of thing are not easily gained by direct observation. For example, iron rust is a familiar chemistry variety. If you don't control reaction conditions, such as water, gas, oxygen, carbon dioxide, the impurity in the air and temperature, etc., it is not easy to understand the reaction and the production.

Regardless of observation or experiment, the fact collected has to be practically accurate. Various operations in the chemistry experiment, such as depositing, filtering, burning, weighing, distilling, titration, crystallization, extraction etc., are all experiments mean to acquire fact knowledge of credible exactitude under the controlled condition. Acquiring exactitude of the knowledge depends on well-trained techniques, and also demands appropriate instruments. Modern chemistry comes of weighing scales application, by measuring each phenomenon, and denoting in numeral, and then you will have accurate knowledge of this phenomenon.

(2) Establishment of Laws

In order to transform the knowledge into scientific theories, you have to collect, analyze, and compare a great deal of the facts by collecting, and inducing from similar facts to get laws.

(3) Founding Theory

Although the number of chemistry laws is less than that of facts, still there are many respective laws which are not correlative with each other. Chemists are required to comprehend the meaning of each law and their correlations. Dalton put forward concept that material is constituted by atoms, and founded atom theory. He explained each law concerning the chemical element with compound variety weight relation, and combined them, thus chemistry knowledge became a systematic science according to formation structure.

There are numerous arising research realms when inorganic chemistry meets other academical areas, for example, interdisciplinary research in the fields of inorganic chemistry and biochemistry brings the inorganic biochemistry.

The application of modern physics experimental methods, such as X ray, neutron diffraction, electronics diffraction, spectrum, mass spectrum, chromatogram etc., leads to a shift of inorganic matter research from macrocosm to microcosm, thus the property of chemical elements and compounds and reactions associated with structure, forms modern inorganic chemistry. Developing trend of inorganic chemistry is mainly synthesis and application of new compounds, and the establishment and development of new research realm.

Organic Chemistry

有 机 化 学

What is organic chemistry?

The word "organic" is one of the most overused words in English world.

People use it as a derogatory term in phrases like "don't eat that"; it's not organic. Of course, there is a precise scientific definition of the word. In science, organic can be a biological or chemical term. In biology, it means anything that is living or has lived. The opposite is non-organic. In chemistry, an organic compound is the one that contains carbon atoms.

Organic chemistry started as the chemistry of life, became the chemistry of carbon compounds, especially for those found in coal, when that was thought to be different from the chemistry in the laboratory. Now it is both. It is the chemistry of the compounds of carbon along with other elements found in living things and elsewhere.

Today, the organic compounds available to us are those presented in living things and those formed over millions of years from dead things. In earlier times, the organic compounds known from nature were those in the "essential oils". Cis-jasmone is a famous example of a perfume distilled from jasmine flowers. In the 20th century, oil overtook coal as the main source of bulk organic compounds so that simple hydrocarbons like methane and propane became available for fuel.

There are about 18 million organic compounds known up to now. How many more are possible? There is no limit.

But these millions of compounds are not just a long list of linear hydrocarbons, they embrace all kinds of molecules with amazingly various properties. They may be crystalline solids, oils, waxes, plastics, elastics, mobile or volatile liquids, or gases. Familiar ones include white crystalline salt, a cheap natural compound isolated from sea water as hard white crystals when it is purificated. Color is not the only characteristic by which we recognize compounds. Generally speaking, it is their odor that let us know when they are around, there are some quite foul organic compounds, too. For example, there are two candidates for this dreadful smell-propane dithiol or 4-methyl-4-sulfanylpentan-2-one. It is unlikely that anyone else will be brave enough to resolve the controversy. Certainly other compounds have delightful odors. Damascenones are responsible for the smell of roses. If you smell one drop you will be disappointed, as it smells rather like turpentine or camphor. But next morning you and your clothes will smell powerfully of roses.

Some organic compounds have strange effects on people. Various drugs such as alcohol and cocaine are taken in various ways to make people temporarily happy. But they have their dangers. If you take too much alcohol, you will suffer from a lot of misery and any cocaine at all may make you a slave for life. One of the latest molecules to be recognized as an anticancer agent in our diet is conjugated linoleic acid in dairy products.

For our third edible molecule, we must choose vitamin C. This is an essential factor in our diets. It is also a universal antioxidant, scavenging for rogue free radicals and protecting us against cancer.

All over the world there are many chemistry-based companies making organic molecules on scales varying from a few kilograms to thousands of tonnes per year. The scale of some of these productions of organic chemistry is almost incredible. The petrochemical industry processes over 10 million liters of crude oil every day.

Some simple compounds are made both from oil and from plants. The ethanol is largely made by the catalytic hydration of ethylene from oil, used as a starting material to make other compounds in industry. The particular acids, bases, surfactants, and so on are chosen to blend together in a smooth emulsion when propelled from the container, and are used to produce soap, detergent, cleaners, and so on, the product should feel, smell, and look attractive and a greenish color is considered clean and antiseptic by the customer.

There is a big market for intense colors for dyeing cloth, coloring plastic and paper, painting walls, and so on. This is the dyestuffs and pigments industry. The most famous dyestuff is probably indigo, an ancient dye that used to be isolated from plants but is now made chemically. It is the color of blue jeans. More modern dyestuffs can be represented by ICI's benzodifuranones, which give fashionable red colors to synthetic fabrics like polyesters.

Many of flavors and fragrances come from oil but others come from plant. A typical perfume will contain 5%~10% fragrances in an ethanol/water (about

90 : 10) mixture. So you might think, the perfumery industry require a very large amount of ethanol and not much perfumery material.

Chemists produce synthetic flavorings such as "smoky bacon" and additive. Meaty flavors come from simple heterocyclic such as alkyl pyrazines and furonol, originally found in pineapples. Compounds such as corylone and maltol give caramel and meaty flavors. Mixtures of these and other synthetic compounds can be tuned to taste like many roasted foods from fresh bread to coffee and barbecued meat.

The pharmaceutical manufacturers produce drugs and medicinal products of many kinds. One of the great revolutions of modern life has been the expectation that humans will survive diseases because of a treatment designed to deal specifically with that disease. The treatment of infectious diseases relies on antibiotics such as the penicillin to prevent bacteria from multiplying. One of the most successful cases of these is Smith Kline Beecham's amoxycillin.

We cannot maintain our present high density of population in the developed world, nor deal with malnutrition in the developing world unless we preserve our food supply from attacks by insects and fungi and from competition by weeds. The world market for agrochemicals is over 10 000 000 000 per annum divided roughly equal among herbicides, fungicides and insecticides.

As you learn more chemistry, you will appreciate how remarkable it is that chemists should produce three-membered rings and use them in bulk compounds to be sprayed on crops in fields. Even more remarkable in some ways is the new generation of fungicides based on a five-membered ring containing three nitrogen atoms—the triazole ring. These compounds inhibit an enzyme present in fungi but not in plants or animals.

All the compounds shown you are built up on hydrocarbon skeletons. Most have oxygen or nitrogen as well; some have sulfur or phosphorus. These are the main elements of organic chemistry. Another definition of organic chemistry would use the periodic table. The key elements in organic chemistry are of course C, H, N, and O, and also include the halogens (F, Cl, Br, I), p-block elements such as Si, S, and P, metals such as Li, Pd, Cu, and Hg, and many more.

So where does inorganic chemistry end and organic chemistry begin? Would you say that the antiviral compound foscarnet was organic? It is a compound of carbon with the formula CPO_5Na_3 but it has no C—H bonds. Does it belong to organic compound or inorganic compound? The answer is that we don't know and we don't care. In fact, it should be noted that distinguishing the strict boundaries between traditional disciplines are undesirable and meaningless. Be glad that the boundaries are indistinct as that means the chemistry is all the richer.

Analytical Chemistry

分析化学

Analytical chemistry is the study on the chemical composition of natural and artificial materials. It consists principally of two major branches—qualitative analysis and quantitative analysis. In practice, quantifying an analyte in a complex sample becomes a complex process. To be efficient and effective, when preparing to analyze a substance of strictly unknown composition, qualitative analysis should precede any attempt at quantitative investigation, because the method of approach selected for the quantitative program may depend on the results of the qualitative analysis.

Broad background knowledge of chemical and physical concepts is required by analytical chemistry. These hypermedia tutorial documents contain links to the fundamental principles that underly the different analytical methods. When studying the analytical chemistry topics, you follow the basic concepts that you are not familiar with or with which you need a refresher. Studying the fundamental concepts will reinforce your understanding of both the analytical techniques and the underlying principles.

Handling the analytical toolbox requires chemists to understand the basic principles of the analytical techniques. With a fundamental understanding of analytical methods, a scientist faced with a difficult analytical problem can apply the most appropriate technique(s). A fundamental understanding also makes it easier to identify when a particular problem cannot be solved by using traditional methods, and gives an analyst the knowledge that it is needed to develop creative approaches or new analytical methods.

The familiar chemistry analytical method include physical means, mass, color, refractive index, thermal conductivity, with electromagnetic radiation (spectroscopy), absorption, emission, scattering, electrochemistry, mass spectrometry.

As you can see, there are a limited number of ways to detect an analyte. However, in each of the above general categories, there are a large number of specific analytical techniques.

1. Separations

A sample that requires analysis is often a mixture of many components in a complex matrix. For samples containing unknown compounds, the components must be separated from each other so that each individual component can be identified by other analytical methods. The separation properties of the components in a mixture are constant under constant conditions, and therefore once determined they can be used to identify and quantify each of the components. Such procedures are typical in chromatographic and electrophoretic analytical separations.

A mixture can be separated by using the differences in physical or chemical properties of the individual components. An analogous example is the filtering of a solid precipitate to separate it from a solution. These separations are based on the states of matter of the two components. Other physical properties that are useful for separations are density and size. Some useful chemical properties by which compounds can be separated are solubility, boiling point, and vapor pressure.

2. Gravimetry

Gravimetry is the quantitative measurement of an analyte by weighing a pure, solid form of the analyte. It is one of the oldest techniques of quantitative analysis. In gravimetry, the analyte is precipitated as a stoichiometrically defined compound. After collecting and drying, the precipitate is weighed on an analytical balance and the analyte is determined from the mass and the known stoichiometry of the analytes in the precipitated compound. Although many instrumental methods have superseded gravimetry, it is still a very important method for standardization processes. Since gravimetric analysis is an absolute measurement, it is the principal method for analyzing and preparing primary standards.

The typical experimental procedure to determine an unknown concentration of an analyte in solution is as follows:

①quantitatively precipitate the analyte from solution

②collect the precipitate by filtering and washing it to remove impurities

③dry the solid in an oven to remove solvent
④weigh the solid on an analytical balance
⑤calculate the analyte concentration in the original solution based on the weight of the precipitate

3. Chemical Equilibrium

In stoichiometry calculations, we assume that chemical reactions would completely reacted and all the reactants or precursors were all consumed at the end of the chemical process. However, when a chemical reaction is carried out in a closed vessel, the system achieves equilibrium. Equilibrium occurs when there is a constant ratio between the concentration of the reactants and the products. Different reactions have different equilibriums. Chemical species will always exist in equilibrium with other forms of itself. The other forms may exist in undetectable amounts but they are always present. For example, pure water consists of the molecular compound and dissociated ions that exist together in equilibrium:

$$H_2O_{(l)} \rightleftharpoons H^+_{(aq)} + OH^-_{(aq)}$$

The (l) subscript refers to the liquid state, and the (aq) subscript refers to ions in aqueous solution.

The equilibrium between reactants and products is described by equilibrium constant. This can be expressed by concentrations of the products divided by the concentration of the reactants with the coefficients of each equation acting as exponents. It is important to remember that only species in either the gas or aqueous phases are included in this expression because the concentrations for liquids and solids cannot change. For the reaction:

$$jA + kB \longrightarrow lC + mD$$

the equilibrium constant K_{eq} is defined as:

$$K_{eq} = \frac{[C]^l[D]^m}{[A]^j[B]^k}$$

where the [ ] brackets indicate the concentration of the chemical species.

Take water for example, $H_2O \rightleftharpoons H^+ + OH^-$, the equilibrium constant is:

$$K_{eq} = \frac{[H^+]^c[OH^-]^d}{[H_2O]}$$

The concentration of water in a water solution is constant and this expression simplifies to:

$$K_w = (55.56 \text{ M}) \times K_{eq} = [H^+][OH^-]$$

where K_w is called the dissociation constant of water and equals to 1.00×10^{-14} at room temperature. The concentrations of [H^+] and [OH^-] therefore equal to 1.00×10^{-7} M.

4. Titration

Titration is the quantitative measurement of an analyte in solution by completely reacting it with a reagent.

The point at which all of the analyte is consumed is called the endpoint and is determined by some type of indicator that is also present in the solution. For acid-base titrations, indicators are available that change color when the pH changes. When all of the analyte is neutralized, further addition of the titrant inducing the pH of the solution to change eventually causes the color of the indicator to change.

The analyte concentration is calculated from the reaction stoichiometry and the amount of reagent that was required to react with all of the analyte.

Manual titration is done with a buret, which is a long graduated tube to hold the titrant. The amount of titrant used in the titration can be read from the difference volume of titrant in the buret between the beginning and end states during the titration process. The most important factor for making accurate titrations is to read the buret volumes reproducibly. The figure shows how to do so by using the bottom of the meniscus to read the reagent volume in the buret.

For repetitive titrations, autotitrators with microprocessors are available to deliver the titrant, stop at the endpoint, and calculate the concentration of the analyte. The endpoint is usually detected by some type of measurements.

5. Electroanalytical Chemistry

Electroanalytical chemistry is a science that makes use of material electricity and electrochemistry property carries on token and measure, and it is also an important compositive part of electrochemistry and analytical chemistry.

A good working definition of the field of electroanalytical chemistry is that it is the field of electrochemistry that utilizes the relationship between chemical phenomena which involve charge transfer and the electrical properties that accompany these phenomena for some analytical determination. This relationship is further broken down into fields based on the type of measurement that is made.

Research contents mainly include:

(1) Composition and appearance analysis.

(2) The electrode process dynamics and electrode reaction mechanism analysis.

(3) Surface and interface analysis.

Potentiometry involves the measurement of potential for quantitative analysis, and electrolytic electrochemical phenomena involve the application of a potential or current to drive a chemical phenomenon, resulting in some measurable signal which may be used in an analytical determination.

6. Spectroscopy

Spectroscopy is the use of the emission, absorption, or scattering of electromagnetic radiation by atoms or molecules (or atomic or molecular ions) to qualitatively or quantitatively study the atoms or molecules, or to study physical processes. The interaction of radiation with matter can cause redirection of the radiation and/or transitions between the energy levels of the atoms or molecules. A transition from a lower level to a higher level with transfer of energy from the radiation field to the atom or molecule is called absorption. A transition from a higher level to lower level is called emission if energy is transferred to the radiation field, or nonradiative decay if no radiation is emitted. Redirection of light due to its interaction with matter is called scattering, and may or may not occur with transfer of energy, i.e., the scattered radiation has a slightly different or the same wavelength.

7. Emission

Atoms or molecules that are excited to high energy levels can decay to lower levels by emitting radiation (emission or luminescence). For atoms excited by a high-temperature energy source, this light emission is commonly called atomic or optical emission, and for atoms excited with light, it is called atomic fluorescence. For molecules, it is called fluorescence if the transition is between states of the same spin and phosphorescence if the transition occurs among states of different spin.

The emission intensity of an emitting substance is linearly proportional to analyte concentration at low concentrations, and is useful for quantitating emitting species.

8. Absorption

When atoms or molecules absorb light, the incoming energy excites a quantized structure to a higher energy level. The type of excitation depends on the wavelength of the light. Electrons are promoted to higher orbitals by ultraviolet or visible light, vibrations are excited by infrared light, and rotations are excited by microwaves.

An absorption spectrum is the absorption of light as a function of wavelength. The spectrum of an atom or molecule depends on its energy level structure, and absorption spectra are useful for identifying compounds.

As long as materials have absorption spectrum, we can carry on qualitative analysis and quantitative analysis. Measuring the concentration of an absorbing species in a sample is accomplished by applying the Beer-Lambert Law.

9. Scattering

When electromagnetic radiation passes through matter, most of the radiation continues in its original direction, but a small fraction is scattered in other directions. Light that is scattered at the same wavelength as the incoming light is called Rayleigh scattering. Light that is scattered in transparent solids due to vibrations (phonons) is called Brillouin scattering. Brillouin scattering is typically shifted by 0.1 cm^{-1} to 1 cm^{-1} from the incident light. Light that is scattered due to vibrations in molecules or optical phonons in solids is called Raman scattering. Raman scattered light is shifted by as much as $4\ 000$ cm^{-1} from the incident light.

10. Data Handling

Since analytical chemistry is the science of making quantitative measurements, it is important to process the raw data correctly in order to give a realistic estimate of the uncertainty in a result.

Simple data processing may only require keeping track of significant figures. More complicated calculations require propagation of error methods.

The uncertainty in a result can be categorized into random error and systematic error.

See the statistical formula for more quantitative descriptions of describing and testing data sets.

11. Analytical Standards

When different methods are adopted to measure the same sample, a problem arises that these results can't be used to compare, so it is necessary to carry out standard analysis method. The standard is the result of standardizing activity, standardizing is the work that has height policy, economy, technique, strictness with continuity, and it is very necessary to set up strict organization to carry on this project.

Standards are materials containing a known concentration of an analyte. They provide a reference to determine unknown concentrations or to calibrate analytical instruments.

The accuracy of an analytical measurement depends on how close a result comes to the true value. Determining the accuracy of a measurement usually requires calibration of the analytical method with a known standard. This is often done with standards of several concentrations to make a calibration or working curve.

Physical Chemistry

物 理 化 学

Physics and chemistry are two branches of natural science with strong correlation with each other, any chemistry variety always goes with physics variety, function of physics factors will also cause chemistry variety. In the natural science, chemistry and physics are as close as brothers, complementing each other. Although they once were established by usage of division, each attends to their own duties, and work respectively. In the natural science, chemistry and physics match together to form an axis. Chemist and physicist carry on with cooperation, promotion in the history, research of many scientists include physics and chemistry. Whenever the chemists tried to make explanation towards laboratory finding, and wanted to refine them into theories, whenever they meet obstacle in the research, they always asked for help from physics achievement at that time, and fortunately benefited a lot. People take notice of contact of physics and chemistry in the long-term of the fulfillment, and take into summary, gradually, an independent academics branch forms—physical chemistry. Since 20th century, the emergence of modern physics greatly promotes the development of chemistry, and the interdisciplinary studies have been made much more efforts. In nowadays, physical chemistry has already been an important academic area in the natural science.

1. Research Contents of Physical Chemistry

Physical Chemistry is the science of basic regulation to investigate chemistry variety from material contact of physicochemical phenomena. The main mission of the physical chemistry is to study the following aspects:

(1) In the appointed condition, whether a chemical reaction automatically react or not, which direction it is carried on, what extent it can arrive, how much energy variety it has in the reaction, how to influence reaction of direction and limit by external condition. The research of these problems

belongs to a branch of physical chemistry, called chemistry thermodynamics.

(2) How fast a chemical reaction is, how to influence reaction velocity by the external environmental condition, which concrete step in a complications the reaction falls into, belong to another branch of physical chemistry, called chemistry kinetics.

(3) Understanding chemical system of microcosmic structure, research regulation of how the atoms combine to molecule in the space member. Because essential property of material depends on its inner structure, through understanding of material inner structure, it is helpful to comprehend inside cause of chemistry variety, and can be helpful to foresee what kind of variety will occur at appropriate of outside function. This belongs to another branch of physical chemistry, called material structure. Material structure research includes structure chemistry and quantum chemistry.

The existing academics is artificially divided, natural science have only six stair academics: astronomy, geography (geology), biology, mathematics, physics, chemistry, the academics cross gradually form a batch crossed academics, intercross of chemistry and physics form physical chemistry and chemical physics.

Physical chemistry is the academics which lies in the middle of physics and chemistry. Due to the theoretical way of its foundation, it is also called theoretical chemistry. And it belongs to a branch of chemistry (a second class academics), which is a basic academics.

The relation between physical chemistry and medicine is very close, for pharmaceutical profession, physical chemistry is a basic specialized course.

2. Formation, Development and Prospect of Physical Chemistry

(1) Formation of physical chemistry

The foundation of physical chemistry can trace back to more than 100 years ago, and the phrase "physical chemistry" was put forward in 1750s. Russian chemist and physicist M. B. Ломоносоb put forward the concept of physical chemistry: "The physical chemistry is an academics, which is based on the principle of the physics and experiment to elucidate material complications after chemistry processing". At that time, he also opened the course, and gave lectures on physical chemistry to his students. A long time after that, although many scientists had done a lot of researches in physical chemistry, physical chemistry didn't form a real academics. In 19th century, large-scale industry production pushed each academics of natural science to a fast development. In the meantime, atom-molecular theory, air molecular movement theory and chemical element periodic law had been already established, a great deal of experience in the chemistry still demanded further summary, and at this time, physical chemistry began to form.

In 1887, the "father" of physical chemistry, famous German chemist W. Ostwald and Dutch chemist Van't Hoff established German Physical Chemistry in common, enabling physical chemistry to be a real independent academical realm, where in the first issue of *Physical Chemistry*, "ionization theory" of Swedish chemist S. A. Arrhenius was announced. All of these three people are important founders of physical chemistry, due to their outstanding contribution and research work to physical chemistry. Therefore, they were usually called "three swordsmen of physical chemistry".

From physical chemistry research in 19th century, physical chemistry formation resulted from much more scientists' long-term hard work, in the process of formation and the earlier research, cross permeation of chemistry and physics was important, the scientists' scientific spirit and research methods, mode of thinking should still be respected.

(2) Development and prospect of physical chemistry

Physical chemistry developed very quickly in 20th century, obtained many great results. According to the statistics, among the Nobel Chemistry Prize winners in 20th century, about 60% of them were engaged in physical chemistry research, in Chinese science department, about 1/3 of chemistry academicians are engaged in researching physical chemistry. As a fundamental discipline of chemistry with great vitality, physical chemistry is important for the formation and development of new inter-disciplinary subject. Now, a theoretical system has already been established in physical chemistry, and Table 1 lists main branches of physical chemistry.

Table 1

Chemistry Thermodynamics			Phase Equilibrium
Chemistry Kinetics			Electrochemistry
Material Structure	Surface Chemistry		Structural Chemistry
	Quantum Chemistry		Photochemistry
			Colloid Chemistry

Physical chemistry in the 20th century was like a running horse, the chemical thermodynamics, chemical kinetics, structure chemistry and quantum chemistry were four legs of this horse and constituted the fundamental theoretical system of physical chemistry. Combined with other realms of physics and chemistry, they formed other branches of physical chemistry, such as photochemistry, electrochemistry, colloid chemistry, surface chemistry, etc.. Structure chemistry and quantum chemistry had already developed to be two chemistry subjects, and were conducted as two specialized courses in the chemical area.

Here, we mainly introduce the development of physical chemistry on the basis of fundamental theories.

Chemical thermodynamics: Thermodynamics is a science that deals with energy conversion regulation of heat and work. From the middle of 19th century to 20th century, the first, the second and the third laws of thermodynamics were proposed. An integral theoretical system has already been formed, which was called classic thermodynamics. Although the theories of classic thermodynamics have been mature, they are limited in an isolated or closed system, and are placed in the equilibrium state. However, occurrence of majority process is not placed in equilibrium state and in closed system. For example, because the biosphere is an opening environment, the theories of classic thermodynamics cannot be used to explain problems about life process. According to the classic thermodynamics, the isolated system is inclined to cancel differences and to stay in the equilibrium state by itself, from the ordered to the disordered. But according to the Darwinian Theory of Evolution, living creatures evolved from the unicellular to multicellular, from the simple to the complex, from the disordered to the ordered, which is contradictory with the conclusion in classic thermodynamics. This antinomy is called the Antinomy of Darwinian and Clausius in the history.

Non-equilibrium thermodynamics has solved this antinomy. Non-equilibrium thermodynamics deals with various problems under non-equilibrium condition and in an open system. This part has two major theories, one is reciprocal relations that is put forward by American chemist Onsager, who was born in Norway; the other is dissipative structure theory that is put forward by Belgian chemist Prigogine. This theory pointed out that system inner dissipative energy of nonreversible process can create or maintain time-space order structure under open and non-equilibrium condition, but living creature structure is the kind of structure that contain preface structure. Onsager and Prigogine have respectively won the Nobel Chemistry Prize in 1968 and 1977.

Nowadays, the hotspot of the chemical thermodynamics research includes biological thermodynamics and heat chemistry research, such as heat chemistry research on cell growth process, thermodynamic research on protein fixed-point incise reaction, thermodynamics research on biological film molecule etc. Moreover, non-linear and non-equilibrium chemical thermodynamics and chemical statistics research, molecule-molecule system heat chemistry research (includes molecular force field, molecule and molecule interaction) etc., are also important aspects.

Chemical kinetics: the research of chemical kinetics is on chemical reaction velocity and the mechanism of chemical reaction, such as reaction velocity equation, activating energy etc.. The research of reaction mechanism includes two aspects: one is reaction type and reaction species, the other is an experimental method.

The research of the reaction type: the detection of the chain reaction begins to investigate the kinetics of complicated reactions. Chain reaction is that free radicals participate in the reaction. N. Semenov won the 1956 Nobel Chemistry Prize due to his theory on chain reaction. The research on light reaction has already formed photochemistry science. Light reaction is a kind of reaction caused by sunlight. One of the most widespread light reactions is photo-synthesis in the natural plants. This aspect of research has obtained some breakthroughs. Making use of solar energy also belongs to the research filed of photochemistry. In the pharmaceutical research, the light stability in medicine research has been valued. According to statistics, in the medical dictionaries in China, more than 60% of the medicines are unstable in the light. In the *Medical Dictionary* (2000), there are explicit provisions on medicine light stability examination, but reaction mechanism of medicine light analysis is still not quite explicit. In addition, the catalyst reaction is an important part to study dynamics. Enzymatic catalyst reaction study is closely related to life science and medicine. This study has mainly two purposes: one is to imitate living, then study emulation to synthesize efficiently, with single mind, gentle catalyst, create medicine with special effect; another purpose is to explain biological phenomena. The life activity is a series of chemical reactions, and the study on enzyme enables us to know the essence of biological phenomena. This aspect of study has already made some important achievements, for example, study on synthetical and analytical mechanism of three phosphoric acid adenosine, ATP, molecular with energy in the cells of mankind and animals. And function mechanism of ATP's synthesizing enzyme has already been clarified in detail, and this achievement was given Nobel Chemistry Prize in 1997 and it announced profound mystery of the energy conversion process in the life.

Research on reaction species originates from research on free radical chemistry. Free radical is a kind of atom or gene that takes nondidymous electronics and has very high chemical activity. Oxygen free radical is most closely related to life body, it exists in all aerobic living creatures, which is essential to life activities. Without it, the life can't continue. But, if the oxygen free radical is excessive, its high chemical activity would hurt the body, and cause diseases. Therefore, a new branch of medical science called free radical medical science is on the rise. Great process on the free radical reaction kinetics in free radical chemistry realm has been made in the Chinese famous universities, such as Peking University, Fudan University and University of Science and Technology of China, etc..

The other hand of research on reaction mechanism is research and experiment methods, mainly making use of various fast probing methods to investigate how the chemical reaction is carried on. The achievements of this aspect are a lot, such as the chemistry relaxation technology and flash photolysis technology. They can exam life span within 10^{-6} or 10^{-9}, even 10^{-12} second, thus come to detect concrete step of chemical reaction. The two techniques won the Nobel Chemistry Prize in 1967. After the 1970s, cross molecular bunch technology was used in the research of dynamics reaction mechanism. This technique is that reaction matter has become two bunch molecules, makes them strike with different directions, differentiates the speed, and uses laser to make molecules to reach excited state. Doing research on reaction and outcome after colliding, this technique makes investigations into dynamics at molecule level, and therefore it is called molecule reaction kinetics research. This achievement got the Nobel Chemistry Prize in 1986. One of the three winners is Chinese American Lee Yuan-tseh. He applied laser technique to the research of super quick process and transition state. Flying laser is a kind of super and short laser pulse. And a flying second is a time unit, with 1 fly second equal to 10^{-15} second. Flying laser technique can measure chemical reaction in several flying seconds. Correspondingly, a new discipline called flying chemistry is established. So far, various chemical reaction mechanisms and the life process mechanism have already been treated at the member level. And the research on flying second chemistry got the Nobel Chemistry Prize in 1999.

   Chemical dynamics as a fundamental research subject in chemistry will have a new development in 21th century, such as making use of molecular bunch technique and laser to research state-state reaction kinetics, using stereoscopic chemical dynamics to research size, shape and space of reaction molecule which influence reaction activity and velocity in the process of the reaction, and using flying laser to study chemical reaction and control chemical reaction process, etc.

   Structure chemistry: Structure chemistry is a science that does research on the structure of molecules and crystals. Currently, structure chemistry has been developed from purely clarifying molecule structure to researching material surface structure, inner structure, dynamic state structure, etc.. Structure analysis can ask for modern spectrum technique and diffraction analysis to carry on, the most direct measurement is a crystal structure analysis. It can be divided into 2 types, namely, X-radial diffraction analysis and microstructure method. The various microstructure techniques will provide emollient weapons for structure chemists, and investigate structure and variety of the big biological molecules, cell, solid surface, etc.. Crystal structure of the big biological molecules, namely, the protein crystal science which provides a condition for the occurrence and development of molecular biology. The 1982 Nobel Chemistry Prize winner, A. Klug founded "the crystal electronics microstructure science" and used it to work out the compound structure of the nucleic acid-protein. This kind of 3D rebuilding technique improved the visual field of electronic microscope from two dimension space to 3D space. A. M. Cormack invented X-radial broken layer diagnosis instrument (CT) for medical diagnosis, and acquired the 1979 Nobel Physiology or Medical Science Prize. In summary, with the development of analyzing instruments and measurement accuracy, new structure analysis is put forth continuously, and the structure chemistry will achieve its ambition in 21th century. The structure research on big biological molecules in the past mainly depended on the X-crystal structure analysis. Since all of them with function are actually in the solution, and their structure is inconstant, after 20th century, nuclear magnetic resonance method is used in the research of big molecules to cause people in the dynamic state structure in the aqua. Catalyst research promoted surface structure research, using STM or AFM and other spectrums to catalyse surface structure. What is more, the catalysing process has also made significant achievements.

Quantum chemistry: The quantum chemistry is a science that studies chemical bonds. Based on the basic theories—Schrödinger equation of the quantum mechanics, we can calculate molecule structure and property. In the 20th century, quantum mechanics and chemistry were combined together. It is very important to understand the chemistry key theories and material structure. Quantum chemistry has already become universal means in the explanation and estimate of molecule structure and chemical behavior. The quantum chemistry once took the chemistry into a modern era in the 20th century. In this modern era, experiments and the theories can join together in the study on the property of molecule system. Since 1928, L. C. Pauling put forward Valence Bond Theory, R. S. Mulliken proposed Molecular Orbit Theory, and H. A. Bethe put forth the Ligand Field Theory. The Molecular Orbit Theory was put forward by R. B. Woodward and R. Hoffmann, the Frontier Orbit Theory was proposed by Fujii, 1998 Nobel Chemistry Prize winner W. Kohn proposed Electronics Density Theory and J. A. Pople raised calculation method of the quantum chemistry and model chemistry. The development has lasted exactly for 70 years. The historical process takes a wide view of the development of quantum chemistry. It is not difficult to see that, only by combining the basic principle of the quantum mechanics and chemistry experiment, the research of quantum chemistry theories can achieve continuously new breakthroughs. Now, based on the quantum chemistry calculation, we can carry on reasonable molecular design, such as medicine design, material design, property estimate, etc.. Someone has foreseen that quantum chemistry as foundation can solve and explain all problems in chemistry experiments in the 20th century.

In summary, the development of physical chemistry is both a breakthrough and a amalgamation. The breakthrough is to investigate new science regulations and technology, such as from the classic equilibrium thermodynamics to non-equilibrium thermodynamics, and from macroscopical reaction kinetics to microcosmic reaction kinetics. And the amalgamation is the interaction and integration of different subjects together to work out various problems, forming edge academics, above various great achievements, all is result that many academics research in common.

Macromolecular Chemistry

大分子化学

Macromolecular chemistry is a integrative science of the physical, biological and chemical structure, properties, composition, mechanisms, and reactions of macromolecules. The history of macromolecular synthesis is just 80 years, and macromolecular chemistry became a real science about 60 years ago. However, it has been developing very fast. Currently, the contents have already gone beyond chemistry scope. Therefore, now people often use macromolecular science to logically address this academic. Macromolecular chemistry, in a narrow sense, refers to macromolecular synthesis chemical and macromolecular chemical reaction.

A macromolecule is a molecule that consists of repeated "building blocks" which may not be identical. When the same building block is repeated, this block is called the monomer, and the resulting macromolecule is called polymer. Nylon as an example of a macromolecule is an artificial polymer. Technically speaking, a macromolecule is a polymer. And as far as macromolecules are involved, macromolecules could not dissolved in a solvent, thus the treatment of macromolecules is identical to polymers. However, since the difference between solid state macromolecules or polymers as they are more widely referred to and polymers in solution are huge, when it comes to the behavior and functionality, the IUPAC in the 1980's and specifically during the IUPAC's conference in Rumania mentioned the difference between the two synonyms as being based on solid state and solution, with one overlap being recognized: gels.

In industry, the value of synthetic macromolecules, particularly the polymers and plastics, has risen enormously over the last 60 years. They have made it possible to mould shapes for the first time. When they were first developed, their resistance to rupture and degradation were regarded as profound advantages, but nowadays we seek more biologically degradable plastics such as polyethyleneglycol that pollutes into the environment less.

Biological macromolecules include enzymes, nucleic acids and polysaccharides such as cellulose and starch. Understanding the basic behavior of polymers has enhanced our knowledge of these biological molecules, and studies of partially charged polyelectrolytes have led to a deeper understanding of their biological function. The three-dimensional structure of these large molecules has led to the identification of specific regions that perform specialized activities. A good example is the catalytic role of particular amino acid residues in polypeptide enzymes and the role of functional groups such as biotin or riboflavin in cellular metabolism. The "folding" of macromolecules is now a topic of many scientific investigations since the correct folding of these polymers is critical for normal function. Abnormal folding of particular proteins is the cause of several diseases, including Alzheimer's and Creutzfeldt-Jakob disease.

Macromolecules have an atomic weight larger than most simple molecules; in general, they have an atomic weight of more than 1 000 Daltons. The atomic weight of some can be as high as millions of Daltons. Their mass and size often cause them to break up into smaller pieces simply due to the shear resistance experienced by the macromolecule in a solvent. Any macromolecule keeping in a simple unfolded stretched filament configuration is experiencing stress shear due to its length and overall mass. For that reason, folding is a practical way for macromolecules to avoid these excessive forces. Some macromolecules in solvent fold into an alpha-helix shape, while some fold into more complex folding structures, but all depend on the possible biological functions. These molecules should exert while in a cellular environment. Complex molecules like DNA, RNA and many proteins also use coiling, folding and super-coiling to regain a functional three dimensional conformation. Often the basic backbones of these long molecules are hidden, but in terms of stability and functionality, in essence, their own existence and safety, are ensured. For practical research, artificial macromolecules are used with a mean molecular weight up to 5 million Daltons with a small spread in weight—to have a sample suitable for research. Long and heavy molecules tend to get disrupted, leading to smaller chains that can influence the measurements, and making it hard to understand results in an unambiguous way. Because of these shear breaks, the average molecular weight of a macromolecular sample will have a deviation around the mean value.

The mechanism of supercoiling and folding is presently being researched in detail as its biological function and failure therein have reportedly been connected to diseases like Alzheimer's and Creutzfeldt-Jakob disease. DNA and other large biological molecules seem to have another mechanism similar to the so-called

Khohlov-implosion of macromolecules, a kind of implosive diminishing of the size of the dissolved molecule where its hydrodynamic radius diminishes by a factor of 1 000 or more. This implosion seems to be based on specific environmental conditions including charge effects and low concentrations.

Macromolecules are used for structure materials, replacing the application of timber, metals, porcelain and glass represent good effect. In agriculture, industry and daily life, they have a lot of advantages, such as light weight, antisepsis, gorgeous color, etc., and are often used for machine parts, ship material, pipe, container, agricultural thin film, furniture, children toy, etc., consumedly making people's lives abundant and beautiful.

However, macromolecules also have a few weakness, for example, flammability. If macromolecules were used a lot, you would have to watch out the fire. Thus you have to make macromolecules apyrous so that it can be used in safety.

Moreover, when a great deal of macromolecule materials are used, macromolecules that are discarded as garbage will become a big social problem of pollution. So we have to make sure that macromolecules will disappear after using.

Material Chemistry

材 料 化 学

Material chemistry is a science that studies the design, preparation, composition, structure, characterization, properties and applications of materials from a chemical point of view. It is not only an important branch of material science, but also an integral part of chemistry. It possesses the characteristics of cross discipline and marginal discipline. With the rapid development of the national economy and the continuous development of material science and chemical science, material chemistry has developed rapidly. In the field of the discovery and synthesis of new materials, the development of the preparation and modification of nanomaterials and the innovation of the characterization methods, material chemistry made a unique contribution. The great strategic significance that design new materials at the atomic and molecular level has broad application prospects.

The current material chemistry develops rapidly in varieties of new ceramics (including superconducting materials), magnetic materials, molecular electronics materials, functional polymer materials, film materials, metal and alloy materials, nonlinear optical materials and luminescent materials, which are also the hotspot of application research.

The wide application of materials is the main driving force for the development of materials chemistry and technology. The material with superior performance in the laboratory does not equal to be applied in the actual working conditions, it must be judged by research application, and then take effective measures to improve.

Application research of materials is the basis of failure analysis of mechanical components and electronic components. Through the application of the material, we can find the regularity in materials, so as to guide the improvement and development of materials. The chemical engineering developed along two main lines: on one hand, after induction, integration, formed discipline basic theory involving transport and reaction as its main contents; on the other hand, with the object of service and applications continuing expansion, discipline basic theory and application field continue to produce new growth point and new branch of science, especially with the emergence of new industries, new energy, new materials, biotechnology, chemical engineering plays a significant role in these new areas, also continue to promote its theory and improve the technical level, and hatch material chemical engineering, biochemical engineering, resource chemical engineering, environmental chemical engineering, and other disciplines branch, which has brought new vitality and development space for the development of chemical engineering, material chemical engineering is one of the fastest growing points. It has become one of the hot research fields of modern chemical engineering.

1. Research Method

The design, fabrication and characterization of materials are three important aspects in the materials chemistry research. The design of the material should be based on the analysis of the material performance and the understanding of the material structure, the material actual effect must also be analyzed through the material structure and performance testing, so the characterization of the structure and properties of materials play a very important role in the material research. The characterization of the structure and properties of materials includes the measurement and characterization of material properties, microstructure and composition. The analysis methods of materials can be divided into two categories: classical chemical analysis and instrumental analysis. The former mainly uses chemical methods to achieve the purpose of analysis, the latter mainly use chemical and physical methods to obtain the results, some of the methods used in this kind of analysis should be applied to more complex and specific instruments. Modern analytical instruments are developing rapidly, and most of the analysis work is done by instrumental analysis, but the classical chemical analysis methods are still of great significance. Theinstrumental analysis and chemical analysis are complementary to each other; it is difficult to completely replace another method.

According to the unique chemical properties of various elements and their compounds, the classical chemical analysis is used to make qualitative or quantitative analysis. According to the relationship between the physical and chemical properties of the molecules, atoms, ions or compounds in the material under test, the instrument analysis is used to analyze the matter qualitatively or quantitatively. The main experimental methods for the analysis of material microstructure and chemical composition are the diffraction (X ray diffraction, electron diffraction, neutron diffraction, γ ray diffraction), microscopy (optical microscopy, transmission electron microscopy, scanning electron microscopy, atomic force microscopy) and spectroscopy method (electron probe, Auger electron spectroscopy, X‐ray photoelectron spectrum).

2. Research Progress

(1) Progress in organic polymer materials

The synthesis of phenolic ester opened up the field of polymer science. In the 1930s, the synthesis of polyamide fiber made the concept of polymer been widely recognized. Later, three aspects of the synthesis of polymer, structure and properties research, application maintain and promote each other, so that the polymer chemistry rapidly develop.

Since the advent of organic polymer materials, people continue to study and explore it. Many organic polymer materials with superior performance and wide application range have been developed. For example, the luminescent organic polymer materials, organic polymer materials with good heat resistance, as well as organic polymer material with hygroscopic effect. Due to the fact that the inorganic piezoelectric materials are brittle and cannot meet the needs of the application, in order to expand the application fields of piezoelectric materials, the polymer piezoelectric composites were studied. This kind of polymer piezoelectric composite material has the advantages of high flexibility, good mechanical properties, easy processing and excellent piezoelectric properties.

(2) Progress in inorganic materials

The traditional inorganic non-metallic materials include cement, glass, ceramics, refractory materials, wear-resistant materials, after nearly 20 years of development of material science, the continuing use of new technology and new technique, new types of ceramic materials, artificial crystal, special glass, nanomaterials and porous materials, inorganic fiber, film material appear and has broad prospects in the application of modern industry, modern national defense and modern life.

Nanotechnology, using individual atoms and molecules to create new material technology, are currently listed as key research field in western countries. For example, the diameter of carbonnanotubes is only 1.4 nm, 50 thousand of these carbon nanotubes lined up in a row was the equivalent of thickness of a hair; the strength is 100 times than that of steel. Nanotechnology, as a new technology, has a significant position in national defense, such as nanosilicon based ceramic powder, the powder coated on the plane, which cannot be detected by the radar, only a few countries in the world possess this technology.

Recently, considering from densification to control grain growth, the crystal and the doping selection, the transparent ceramics have important application prospect in lighting, medical equipment, laser, armor and infrared device.

Since 1970s, the research and development of hydrogen energy have become more and more important. The study of hydrogen storage materials is one of the hot spots. Hydrogen storage material is divided into chemical hydrogen storage (such as hydrogen storage alloy, coordination hydride, amino compound, organic liquid, etc.) and physical hydrogen storage (such as carbon based materials, metal organic frameworks, etc.). Some hydrogen storage materials and technologies currently have made important progress in some aspects, however, whether the hydrogen storage density, temperature, cycle performance, or safety, there is a great distance from the practical application of hydrogen energy.

Computational Chemistry

计 算 化 学

In the recent decades, computational chemistry, it is safe to say, is one of chemistry realms which witnessed the quickest development. And what is computational chemistry? Explained in a simple way, computational chemistry is a science that studies the systematic properties of chemistry with a great number of numerical operations according to basic theories in physical chemistry (usually quantum chemistry). The most familiar example is to explain experiments in various chemical phenomena through quantum chemistry calculation, helping chemists to understand and analyse observation results with more concrete concepts. In addition, for unknown or unclear chemical systems, computational chemistry also plays the role of a predictor, providing further research directions. Moreover, computational chemistry is also used frequently to validate, test, revise, or develop theories in higher chemistry.

Many universities are now setting up classes, which are an overview of various aspects of computational chemistry. Since we have had many people hoping to start computations even before their first introductory course, this chapter can be deemed as the first step in understanding what computational chemistry is. This does not intend to teach the fundamentals of chemistry, quantum mechanics or mathematics. It only aims at presenting a basic description of how chemical computations are carried out.

The term "theoretical chemistry" is defined as the mathematical description of chemistry. The term "computational chemistry" is usually used when a mathematical method is so well developed that it can be implemented automatically on a computer. The words, like "exact" and "perfect", do not appear in these definitions. Very few aspects of chemistry can be computed exactly, but almost every aspect of chemistry has been described in a qualitative or approximately quantitative computational scheme. The biggest mistake that a computational chemist can make is assuming that every computed number is exact. However, just as not all spectra are perfectly resolved, a qualitative or approximate computation can usually give useful insight into chemistry if you understand what it tells you and what not.

1. Ab Initio

Over the past three decades, ab initio quantum chemistry has become an essential tool in the study of atoms and molecules and increasingly in modeling complex systems such as those arising in biology and materials science. And then what is ab initio computational method? The term "Ab Initio" is Latin for "from the beginning". This name is given to computations which are derived directly from theoretical principles, with no inclusion of experimental data. Most of the time, this is referred to as an approximate quantum mechanical calculation. The approximations made are usually mathematical approximations, such as using a simpler functional form for a function or getting an approximate solution to a differential equation.

The most common type of ab initio calculation is called a Hartree Fock calculation (abbreviated as HF), in which the primary approximation is called the central field approximation. This means that the Coulombic electron-electron repulsion is not specifically taken into account. However, its net effect is included in the calculation. This is a variational calculation, which means that the approximate energies calculated are all equal to or greater than the exact energy. The energies calculated are usually in unit called Hartrees (1 H = 27.2114 eV). Because of the central field approximation, the energies from HF calculations are always greater than the exact energy and tend to be close to a limiting value called the Hartree Fock limit.

The second approximation in HF calculations is that the wave function must be described by some functional forms, which are only known exactly for a few one electron systems. The functions used most frequently are linear combinations of Slater type orbitals $\exp(-ax)$ or Gaussian type orbitals $\exp(-ax^2)$, abbreviated STO and GTO. The wave function is formed from linear combinations of atomic orbitals or more often from linear combinations of basis functions. Because of this approximation, most HF calculations give a computed energy greater than the Hartree Fock limit. The exact set of basis functions used is often specified by an abbreviation, such as STO-2G or 6-31++g**.

A number of types of calculations begin with a HF calculation, and then are corrected for the explicit electron-electron repulsion, referred to as correlation.

Among these methods are Mohlar-Plesset perturbation theory (MPn, where "n" is the order of correction), the Generalized Valence Bond (GVB) method, Multi-Configurations Self Consistent Field (MCSCF), Configuration Interaction (CI) and Coupled Cluster theory (CC). As a group, these methods named as correlated calculations.

A method, which avoids the HF mistakes in the first place, is called Quantum Monte Carlo (QMC). There are several flavors of QMC, such as variational, diffusion and Green's functions. These methods work with an explicitly correlated wave function and a Monte Carlo integration is used to evaluate integrals numerically. These calculations can be very time-consuming, but they are probably the most accurate methods known today.

An alternative of ab initio method is Density Functional Theory (DFT), in which the total energy is expressed in terms of the total electron density, rather than the wave function. In this type of calculation, there is an approximate Hamiltonian and an approximate expression for the total electron density.

The advantage of ab initio methods is that they eventually converge to the exact solution as long as all of the approximations are made sufficiently small in magnitude. However, this is not the single convergence. Sometimes, the smallest calculation gives the best result for a given property.

The disadvantage of ab initio methods is that they are expensive. These methods often take enormous amounts of CPU time, memory and disk space. The HF method scales as N^4, where "N" is the number of basis functions, so a calculation twice as big takes 16 times as long to complete. Correlated calculations often scale much worse than this. In practice, extremely accurate solutions are only obtainable when the molecule contains half a dozen electrons or less.

In general, ab initio calculations give very good qualitative results and can give increasingly accurate quantitative results as the molecules in question become smaller.

2. Semiempirical

Semiempirical methods are simplified versions of Hartree-Fock theory using empirical corrections in order to improve performance. These methods are usually referred to encode some of the underlying theoretical assumptions through acronyms. The most frequently used methods (MNDO, AM1, PM3) are all based on the Neglect of Differential Diatomic Overlap (NDDO) integral approximation, while older methods use simpler integral schemes such as CNDO and INDO. All of the three approaches belong to the class of Zero Differential Overlap (ZDO) methods, in which all two-electron integrals involving two-center charge distributions are neglected. A number of additional approximations are made to speed up calculations and a number of parameterized corrections are made in order to correct for the approximate quantum mechanical model. How the parameterization is performed characterizes the particular semiempirical method. For MNDO, AM1, and PM3, the parameterization is performed in such a way that the calculated energies are expressed as heats of formations instead of total energies.

The advantage of semiempirical calculations is that they are much faster than the ab initio calculations.

The disadvantage of semiempirical calculations is that the results can be erratic. If the molecule being computed is similar to molecules in the data base used to parameterize the method, then the results may be very satisfactory. If the molecule being computed is significantly different from anything in the parameterization set, the answers may be very poor.

Semiempirical calculations have been very successful in the description of organic chemistry, where there are only a few elements used extensively and the molecules are of moderate size. However, semiempirical methods have been devised specifically for the description of inorganic chemistry as well.

3. Molecular Dynamics

Molecular dynamics is computer simulation of physical movements of atoms and molecules, and it is frequently used in the study of proteins and biomolecules, as well as in material science. This method uses the Newtonian equations of motion, a potential energy function and associated force field to follow the displacement of atoms in a molecule over a certain period of time, at a certain temperature and a certain pressure. Calculations of motion are done at discrete and small time intervals and a velocity calculated on each atom position which in turn is used to calculate the acceleration for the next step. Starting velocities can be calculated at random or by scaling the initial forces on the atoms. Simulations can also be run with differing temperatures to obtain different families of conformers. At higher temperatures, more conformers are possible and it becomes feasible to cross energy barriers.

Molecular dynamics consists of examining the time dependent behavior of a molecule, such as vibrational motion or Brownian motion. This is most often done within a classical mechanical description similar to a molecular mechanics calculation.

For biological molecules, the calculations are more frequently to take the presence of solvent into account. However, this brings further complications due to two main problems. The first is the increasing CPU time due to the larger number of atoms. The second is that the water molecules surrounding the molecule tend to drift away from the molecule of interest and get "lost" from the calculation if only a certain area of space is monitored as in usual cases. This causes nasty "edge effects". There is one method currently used to get around this problem. That is to place your molecule surrounded in water in a box of a specific size and then to surround that box with an image of itself in all directions. The solute in the box of interest interacts only with its nearest neighbour images. Since each box is an image of the other, then when a molecule leaves a box, its image enters from the opposite box and replaces it so that there is conservation of the total number of molecules and atoms in the box. This is known as periodic boundary conditions.

In order to analyze the vibrations of a single molecule, many dynamics steps are done, and then the data is transformed into the frequency domain by Fourier transformation. A given peak can be chosen and transformed back to the time domain, in order to see what the motion at that frequency looks like.

Nowadays, the major software for MD simulations are Auto Dock, AMBER, BOSS, ChemSketch, etc..

4. Statistical Mechanics

Statistical mechanics is the first fundamental physical theory in which probabilistic concepts and probabilistic explanation play a fundamental role. For the philosophers, it provides a crucial test case to compare the philosophers' ideas about the meaning of probabilistic assertions and the role of probability in explanation of what actually goes on when probability enters a foundational physical theory.

Statistical mechanics is the mathematical means to extrapolate thermodynamic properties of bulk materials from a molecular description of the material. Much of statistical mechanics is still at the paper and pencil stage of theory, since the quantum mechanicians cannot solve the Schrödinger equation exactly yet, the statistical mechanicians even does not really have a good starting point for a truly rigorous treatment.

Statistical mechanics computations are often tacked onto the end of ab inito calculations for gas phase properties. For condensed phase properties, molecular dynamics calculations are often needed in order to do a computational experiment.

5. Thermodynamics

Thermodynamics is a branch of physics which deals with the energy and works of a system. It was born in the 19th century as scientists first discovered how to build and operate steam engines. Thermodynamics deals only with the large scale response of a system which we can observe and measure in experiments. Small scale gas interactions are described by the kinetic of gases. The methods complement each other; some principles are more easily understood in terms of thermodynamics and some principles are more easily explained by kinetic theory.

There are three principal laws of thermodynamics which are described on separate slides. Each law leads to the definition of thermodynamic properties which help us to understand and predict the operation of a physical system. We will present some simple examples of these laws and properties for a variety of physical systems, although we are most interested in thermodynamics in the study of propulsion systems and high speed flows. Fortunately, many of the classical examples of thermodynamics involve gas dynamics. Unfortunately, the numeric system for the three laws of thermodynamics is a little confusing.

6. Structure-Property Relationships

Structure-property relationships are qualitative or quantitative empirically defined relationships between molecular structure and observed properties. In some cases, this may duplicate statistical mechanical results. However, structure-property relationships need not to be based on any rigorous theoretical principles.

The simplest case of structure-property relationships are qualitative thumb rules. For example, an experienced polymer chemist may be able to predict whether a polymer will be soft or brittle based on the geometry and bonding of the monomers.

When structure-property relationships are mentioned in current literature, it usually implies a quantitative mathematical relationship. These relationships are often derived by using curve fitting software to find the linear combination of molecular properties, which best reproduces the desired property. The molecular properties are usually obtained from molecular modeling computations. Other molecular descriptors such as molecular weight or topological descriptions are also used.

When the property described is a physical property, such as the boiling point, this is referred to as a Quantitative Structure-Property Relationship (QSPR). When the property being described is a type of biological activity (such as drug activity), this is referred to as a Quantitative Structure-Activity Relationship (QSAR).

At present, in the field of drug research and development, computer-aided design has made tremendous progress. For instance, QSPR/QSAR model has become essential computational tools for drug design.

7. Artifical Intelligence

Techniques invented by computer scientists interested in artificial intelligence have been applied mostly to drug design in recent years. These methods also go by the names like De Novo or rational drug design. The general scenario is that some functional sites have been identified and it is desired to set up a structure for a molecule that will interact with that site in order to hinder its functionality. Instead of having a chemist who would try hundreds or thousands of possibilities with a molecular mechanics program, the molecular mechanics is built into an artificial intelligence program, which tries enormous numbers of "reasonable" possibilities in an automated fashion. The number of techniques for describing the "intelligent" part of this operation is so diverse that it is impossible to make any generalization about how this is implemented in the program.

8. General Computational Research Problem

When computational chemistry is used to answer a chemical question, there is an obvious problem that you need to know how to use the software. The problem that is missed is that you need to know how good the answer is going to be. Here is a check list to follow.

What do you want to know? How accurately? Why? If you can't answer these questions, then you don't even have a research project yet.

How accurate do you predict the answer will be? In analytical chemistry, you do a number of identical measurements then work out the error from a standard deviation. With computational experiments, doing the same thing should always give exactly the same result. The way that you estimate your error is to compare with a number of similar computations to the experimental answers. There are articles and compilations of these studies. If none of them exist, you will have to guess which method should be reasonable based on its assumptions, and then do a study by yourself before you can apply it, in which you have no idea of how good the calculation is. When someone just tells you off the top of their head what method to use, they either have a fair amount of this type of information memorized, or they don't know what they are talking about. Beware of that someone who tells you a given program is good just because it is the only one they know how to use, rather than their answer is based on the quality of the results.

How long do you expect it to take? If the world were perfect, you would tell your PC (voice input of course) to give you the exact solution to the Schrödinger equation and go on with your life. However, ab initio calculations would often be so time-consuming that it would take a decade to do a single calculation if you even had a machine with enough memory and disk space. However, a number of methods exist because each has its most suitable situations. The trick is to determine which one is the best for your project. Again, the answer is to look into the literature and see how long each takes. If the only thing you know is how a calculation scale does the simplest possible calculation, then use the scaling equation to estimate how long it will take to do the sort of calculation that you have predicted will give the desired accuracy.

What approximations are being made? Which are significant? This is how you avoid acting like a complete fool, when you successfully perform a calculation that is complete garbage. An example would be trying to find out vibrational motions that are very anharmonic, when the calculation uses a harmonic oscillator approximation.

Once you have finally answered all of these questions, you are ready to actually do a calculation. Now you must determine what software is available, what it costs and how to use it. Note that two programs of the same type may calculate different properties, so you have to make sure the program does exactly what you want.

Chemical Laboratory

化学实验室

1. Laboratory Safety

When you first come into a chemical laboratory, certain elementary precautions should be taken and each researcher conducts himself with common sense and alertness. Laboratory accidents are often caused by attempts to obtain results in hurry, thoughtless or ignorant behavior. The responsibility to guarantee lab safety rests with each worker and every student in the laboratory. You must use common sense and work carefully to avoid chemical spills, broken glassware, and fires. This ensures not only your own safety, but also that of your lab mates'. You need to know the hazards of each chemical you use so that you will know what level of caution to use when handling it. If you do this, you will not be exposed to a harmful amount of any chemical during your years in a chemistry lab.

If an accident does happen, you must take steps to prevent further injury. Most accidents are minor, and methods of dealing with them are detailed in the sections below. In serious accidents, remember that injured people are often in shock and are unable to help themselves. You should be prepared to help your neighbor in case of an accident. A matter of seconds can be critical.

2. Safety Equipment

Personal safety equipment in the chemistry laboratory includes: first aid kit, fume hoods, safety showers, eye wash stations, fire extinguishers, gas shutoff valve, telephone, safety goggles, gloves, lab coats and aprons and so on.

One first aid kit is located in each laboratory, at the top of the shelf. It contains gauze squares, small, adhesive bandages and antibiotic ointment. If any injury occurs which cannot be handled with these supplies, then the student can be escorted to the hospital to receive treatment from the doctors there, or can wait in the lab for an Emergency Medical Service team. The fume hoods are large cabinets which have sliding glass doors in front. Fume hoods are used to protect you from harmful fumes, gases and odors. The fume hood has an air duct in its ceiling which is attached to a powerful fan. When the fan is turned on, the air in the fume hood is pulled up through the duct, carrying away all harmful fumes or smoke. If chemicals splashed on your face they would not reach your eyes because you are wearing safety goggles. If this sort of accident happens, keep your goggles on while you go to the eye-wash station. There you should wash your face with the goggles still on until you are reasonably sure most of the chemical is gone from your face. Then you should remove your goggles and wash again. There is an emergencies-only telephone in the Chemical Preparations Room. This room is only accessible by Chemistry instructors and staff. If someone asks you to call for help, find a lab instructor, professor or staff member and ask them to call for hospital.

Lab coat is part of your personal safety equipment. You must be covered from the top of the shoulders to well below the knees. Your feet must be covered. Sandals are not appropriate in the chemical lab. Very loose fitting garments, such as ties and wide sleeves, as well as long unrestricted hair pose a hazard and must be restrained.

3. Eye Safety

You must wear chemical spill protection safety goggles whenever anyone in the lab room is handling chemicals. They must be flexible fitting, hood ventilation goggles, according to ANSI chemical splash standards.

Do not wear contacts in the labs, even under goggles. Soft lenses can be affected by solvent vapors, possibly even fusing to the eye. Contacts also make it difficult to rinse your eyes quickly and properly if you spill something in them.

If you got any chemical in your eye, rinsing it immediately in the eye wash, holding your eye open. Your teacher or another student will come to help you ascertain the seriousness of the exposure.

4. Glassware Safety

Use common sense when handling glassware. Keep glassware away from the edge of the benchtop. Always clamp your reaction flask and the suction flask securely to a ring stand to prevent them from falling over. Check each piece of glassware for hairline or star cracks before using it. When doing a distillation, clamp each piece of glassware securely. If you do break a piece of glassware, do not leave it in the sink or on the benchtop because someone may inadvertently get cut. Wearing thick gloves and using a brush and dustpan to sweep up the broken glass. Place the broken glassware in one of the "Broken Clean Glassware" containers located in the labs.

If your reaction use heating mantle or steam bath, the glassware or the clamps used to hold glassware can become hot enough to cause a thermal burn on your skin. Wear heavy gloves to prevent this. Cuts can also be prevented by wearing thick gloves, especially while washing glassware. Protect your feet by wearing closed-toed shoes not only to protect your feet from dropped glassware, but also to protect them from broken pieces of glass which may be on the floor from a previous lab section. Always wear your goggles to protect your eyes from flying broken glassware.

If you cut yourself, wash the wound immediately with large amount of cool water. If it is your neighbor who has been hurt, be prepared to help him if he is unable to help himself. Apply direct pressure to stop the bleeding as necessary. If the bleeding is profuse, elevate the affected limb. Watch for evidence of shock and contact your teacher or the Lab Coordinator if necessary. Thermal burns are treated by covering the affected area with cool water or ice.

5. Special Health Problems

If you are aware that you have allergies to specific chemicals or drugs, or to UV light, or you have asthma or other health problems, you may need to consult your teacher before taking chemistry lab section. Feel free to discuss any questions you may have with the Laboratory Coordinator. And please note: you must not take organic or radioactive chemistry lab if you are pregnant.

6. Hazardous Chemicals

Chemicals in the organic lab can be flammable, volatile, health hazardous, and/or corrosive. In the organic chemistry lab courses, we advise that you know the hazards of all the chemicals in the laboratory. First and foremost, you need to know these hazards so that you will know when it is critical to take precautions such as wearing protective clothing or keeping chemicals from flame. We consider this so important that you will always be asked to look up the hazards and include them in your prelab notebook write-up. And hope that the lessons you learn about the hazards of chemicals will enable you to work in a safe manner whatever your future profession is.

7. General Guide for Handling Chemicals in the Laboratory

Knowing what the hazards are is one thing, and knowing how to handle chemicals with these hazards is another. The chemicals most frequently used in the chemistry laboratories are chosen as examples for each type of hazard.

(1) **Flammable Chemicals** (examples: diethyl ether and methanol)

The method for proper handling these flammable chemicals depends on their flammability rating, as given by a number 4 - 0 in the red area of a National Fire Protection Association (NFPA) label. The NFPA rating for diethyl ether is "4" while acetone, methanol, ethanol, and hexanes are "3". Ether is extremely flammable and any spark or simply heat can ignite it. The other four solvents listed here will readily burn, but they are not as likely to spontaneously combust.

Never use ether in a lab that has an open flame anywhere in the room. Be careful not to spill any flammable solvent (especially ether) on a heating mantle or hot plate.

If your clothing catches fire, immediately drop it to the floor and roll to smother the flames and call for help. If a compound or solvent catches on fire, you can quickly cover the flames with a piece of glassware. If it is feasible, use a fire extinguisher to put the fire out.

Do not put water on an organic chemical fire because it will only spread the fire.

If the fire is large, do not take chances: evacuate the lab and the building immediately and tell your teacher or the coordinator what has happened.

If no one in authority is available, pull the fire alarm in the hallway.

If no one in authority is available, call for 119 from a safe phone.

If the fire alarm sounds for any reason, leave the room immediately and exit the building.

(2) **Volatile Chemicals** (examples: hexanes, acetone, diethyl ether)

Diethyl ether and methylene chloride are the most volatile of the chemicals that you will use in the organic chem teaching labs. If accidentally inhaled, they can cause irritation of the respiratory tract, intoxication, drowsiness, nausea, or even central nervous system depression. Keep in mind that diethyl ether presents a special problem because it is not only volatile, but also extremely flammable.

Work in your student hood whenever possible, especially when you are handling volatile chemicals. If you need to carry volatile chemicals through the lab, do it in a covered container.

Everyone in the lab must work together to reduce the amount of volatile chemicals released into the lab room.

(3) **Health Hazardous Solvents** (examples: methanol, ethanol, diethyl ether)

The health hazard of a chemical is designated by a number 4 – 0 in the blue area of a NFPA label. None of the chemicals you will use has a "4" rating; most are 1 or 2. If you had a one-time overexposure to the above chemicals, you might suffer a minor or a serious injury. If you protect yourself properly by wearing gloves, lab coat, goggles, and closed-toed shoes, and if you are careful not to spill chemicals, you are not likely to come into contact with these chemicals. In the past, chemicals have been spilled by students and left where they were in the lab, especially by the balances. This could cause serious harm to another student, so be sure to clean up a chemical spill promptly. Often in the Lab Manual, you will be directed to clamp reaction flasks and vacuum flasks. The reason for this is two-fold so that you do not lose your product and so that you do not spill chemicals.

(4) **Hazardous Corrosives** (examples: phosphoric acid, sodium hydroxide)

Strong acids and bases are frequently used in the organic chemistry teaching labs. At full strength, they have a health rating of "3", meaning that short exposure could cause serious injury. If spilled on your skin, they would cause a chemical burn. They are very harmful to your eyes. If you breathe in a big whiff of vapors, you will feel a burning in your nasal and respiratory passages.

Handle corrosives with great care so as not to spill them or inhale their vapors. Always wear goggles, gloves, protective clothing, and shoes. The heavier styles of gloves are recommended for use when handling corrosives.

8. The Chemistry Laboratory Notebook

(1) Title and Date

Give the title of the experiment and its date.

(2) Introduction

In a sentence or two, state the purpose of the experiment. If the experiment is a preparative experiment, the introduction should also includes the balanced equation for the reaction.

In chemistry experiment lessons, there are different types of experiments: technique and preparative. A technique experiment is one in which you are performing a technique for the first time and studying its details, for example, distillation and extraction, a preparative experiment is one in which a compound is synthesized from other reagents in organic chemistry experiment.

(3) Physical Data

List the melting point, boiling point, density, solubility, and hazards of all pertinent chemicals used in the experiment. For your convenience, tables of physical data for all chemicals used in this course are included in the Handbook for Chemistry Lab. Or, you can find the information on the Internet.

Calculate the amounts of reactants in moles and grams or mL as applicable. In a preparative experiment, calculate the limiting reagent and the theoretical yield of the product. Be sure to include your calculations for these values. Refer to the Handbook for information on how to calculate yields. The physical data are most conveniently presented in tabular form, although in a preparative experiment you may put the amounts of reactants and products under the balanced equations for the reaction.

(4) Procedure

In the following, briefly summarize the procedure. An outline or a flow chart would be preferred. You do not need to write down the procedure in complete sentences and do not copy directly from the Lab Manual. All you need is a brief but complete listing of what you plan to do in the lab.

(5) **Data and Observations**

Your observations of the experiment as it progresses are important, new information. Write these observations, including color changes, appearance of crystals, formation of an emulsion, boiling temperatures, test results, in your notebook as you do the experiment. Also record the weights of reagents and products and tare weights in this section.

In general, you do not need to re-write the procedure section in these observations, instead, you may state that "the procedure was carried out as planned". At times, however, you may have to write the procedure out partially. For instance, if you state "the solution turned green," you will have to write enough of the procedure so that your teacher will know at what step in the reaction the solution turned green. As a guideline, consider that from the procedure, data and observations sections, any chemist should be able to duplicate your experiment.

(6) **Results and Discussion**

This is the section in which you interpret the data obtained in the previous section. For example, indicate the amount of purified compound that you obtained and how the purity and identity of the compound was assessed. In a preparative experiment, present the percent yield. Include and discuss instrument printouts, such as GC traces and IR spectra. In this section, you can state whether the procedure was a good method for making the desired compound or not; if not, try to make suggestions to improve the method for future experimenters.

Section III Literature Introduction

第三部分 文献导读

本部分主要由四篇文章组成。第一篇文章由美国科学院院士、哈佛大学教授 George M. Whitesides 所写,后面的三篇文章分别取自《The Journal of Chemical Physics》、《Chinese Science Bulletin》和《Catalysis Communications》三种期刊。

George M. Whitesides 教授所写的文章发表在 2004 年的《Advanced Materials》上,主要简要介绍了科技论文写作的基本原则、技巧以及要点。读者可以通过通读全文,了解科技论文写作的基本要点和流程。比如,在课题刚开始的时候如何去构思一个课题的框架,为了撰写一篇严谨的科技论文应该准备哪些数据和材料,怎样设计安排相应的实验。在这篇文章里面 George M. Whitesides 教授将这些内容一一阐述。通过仔细理解这篇文章,读者能够从中得到一些启发,为以后的科技论文写作打下基础。

后面的三篇文章分别涉及无机、有机、催化等方面的内容,是综合性的科技论文,全面地阐述了整个项目的思路和过程。希望读者通过对这三篇文章的学习,对一般的科研项目有一个感官上的认识和理解。另外,这三篇文章涵盖了大部分化学专业相关的常用词汇,认真地学习之后会体会到科技论文写作中的很多共性,对以后的科技论文写作也会有一定的帮助。读者在学习过程中尤其要注意作者对于整个实验的思路构架和对实验结果的探讨,以及用英语对实验过程进行的描述,这些都是值得学习和思考的地方。

为了展现科技期刊的原汁原味,在排版过程中我们特意将能够体现期刊风貌的原文首页以影印形式呈现给读者,意在以直观的方式让读者体会到不同期刊的不同排版设计风格。在投稿的时候不同期刊会一一讲明要求的格式,这里也不做赘述。不过在后面的原文摘录过程中我们有意地体现出原文的风貌,包括参考文献格式,读者可以略见一二。有兴趣的读者也可以自己搜索这些文章来自学体会。

写好科技论文不是一蹴而就的,一方面需要努力学习专业知识,提高专业技术水平、业务和科研能力;另一方面要多阅读相关专业的科技期刊,学习高水平的科研成果和论文写作思路与方法,从中学习专家们严谨的治学态度、先进的科研技术路线与方法、准确严密的表述语言。这是提高自身科技论文写作水平的有效途径。同时,在日常业务、科研工作中,要注意发现问题和积累资料,多分析、多总结、多写作,锻炼自己发现、思考和解决问题的能力以及归纳总结的能力和科技论文的写作能力。

Whitesides' Group: Writing a Paper*

By *George M. Whitesides***

1. What is a scientific paper?

A paper is an organized description of hypotheses, data and conclusions, intended to instruct the reader. Papers are a central part of research. If your research does not generate papers, it might just as well not have been done. "Interesting and unpublished" is equivalent to "non-existent".

Realize that your objective in research is to formulate and test hypotheses, to draw conclusions from these tests, and to teach these conclusions to others. Your objective is not to "collect data".

A paper is not just an archival device for storing a completed research program, it is also a structure for planning your research in progress. If you clearly understand the purpose and form of a paper, it can be immensely useful to you in organizing and conducting your research. A good outline for the paper is also a good plan for the research program. You should write and rewrite these plans/outlines throughout the course of the research. At the beginning, you will have mostly plan; at the end, mostly outline. The continuous effort to understand, analyze, summarize, and reformulate hypotheses on paper will be immensely more efficient for you than a process in which you collect data and only start to organize them when their collection is "complete".

---

[*] The text is based on a handout created on October 4, 1989.
[**] Prof. G. M. Whitesides
Department of Chemistry and Chemical Biology
Harvard University
Cambridge, MA 02138 (USA)
E-mail: gmwhitesides@gmwgroup.harvard.edu

Whitesides' Group: Writing a Paper**

By *George M. Whitesides**

1. What is a Scientific Paper?

A paper is an organized description of hypotheses, data and conclusions, intended to instruct the reader. Papers are a central part of research. If your research does not generate papers, it might just as well not have been done. "Interesting and unpublished" is equivalent to "non-existent".

Realize that your objective in research is to formulate and test hypotheses, to draw conclusions from these tests, and to teach these conclusions to others. Your objective is not to "collect data".

A paper is not just an archival device for storing a completed research program; it is also a structure for *planning* your research in progress. If you clearly understand the purpose and form of a paper, it can be immensely useful to you in *organizing* and conducting your research. A good outline for the paper is also a good plan for the research program. You should write and rewrite these plans/outlines throughout the course of the research. At the beginning, you will have mostly plan; at the end, mostly outline. The continuous effort to understand, analyze, summarize, and reformulate hypotheses on paper will be immensely more efficient for you than a process in which you collect data and only start to organize them when their collection is "complete".

2. Outlines

2.1. The Reason for Outlines

I emphasize the central place of an outline in writing papers, preparing seminars, and planning research. I especially believe that for you, and for me, it is most *efficient* to write papers from outlines. An *outline* is a written plan of the organization of a paper, *including* the data on which it rests. You should, in fact, think of an outline as a carefully organized and presented set of data, with attendant objectives, hypotheses, and conclusions, rather than an outline of text.

An outline itself contains little text. If you and I can agree on the details of the outline (that is, on the data and organization), the supporting text can be assembled fairly easily. If we do *not* agree on the outline, any text is useless. Much of the *time* in writing a paper goes into the text; most of the *thought* goes into the organization of the data and into the analysis. It can be relatively efficient in time to go through several (even many) cycles of an outline before beginning to write text; writing many versions of the full text of a paper is slow.

All writing that I do—papers, reports, proposals (and, of course, slides for seminars)—I do from outlines. I urge you to learn how to use them as well.

2.2. How Should You Construct an Outline?

The classical approach is to start with a blank piece of paper, and write down, in any order, all important ideas that occur to you concerning the paper. Ask yourself the obvious questions: "Why did I do this work?"; "What does it mean?"; "What hypotheses did I mean to test?"; "What ones did I actually test?"; "What were the results? Did the work yield a new method of compound? What?"; "What measurements did I make?"; "What compounds? How were they characterized?". Sketch possible equations, figures, and schemes. It is essential to try to get the major ideas. If you start the research to test one hypothesis, and decide, when you see what you have, that the data really seem to test some other hypothesis better, don't worry. Write them both down, and pick the best combinations of hypotheses, objectives, and data. Often the objectives of a paper when it is finished are different from those used to justify starting the work. Much of good science is opportunistic and revisionist.

When you have written down what you can, start with another piece of paper and try to organize the jumble of the first one. Sort all of your ideas into three major heaps (1–3).

1. Introduction

Why did I do the work? What were the central motivations and hypotheses?

2. Results and Discussion

What were the results? How were compounds made and characterized? What was measured?

3. Conclusions

What does it all mean? What hypotheses were proved or disproved? What did I learn? Why does it make a difference?

---

[*] Prof. G. M. Whitesides
Department of Chemistry and Chemical Biology
Harvard University
Cambridge, MA 02138 (USA)
E-mail: gmwhitesides@gmwgroup.harvard.edu

[**] The text is based on a handout created on October 4, 1989.

2. Outlines

2.1. The reason for outlines.

I emphasize the central place of an outline in writing papers, preparing seminars, and planning research. I especially believe that for you, and for me, it is most *efficient* to write papers from outlines. An *outline* is a written plan of the organization of a paper, *including* the data on which it rests. You should, in fact, think of an outline as a carefully organized and presented set of data, with attendant objectives, hypotheses and conclusions, rather than an outline of text.

An outline itself contains little text. If you and I can agree on the details of the outline (that is, on the data and organization), the supporting text can be assembled fairly easily. If we do *not* agree on the outline, any text is useless. Much of the *time* in writing a paper goes into the text; most of the *thought* goes into the organization of the data and into the analysis. It can be relatively efficient to go through several (even many) cycles of an outline before beginning to write text; writing many versions of the full text of a paper is slow.

All the writing that I do—papers, reports, proposals (and, of course, slides for seminars)—I do from outlines. I urge you to learn how to use them as well.

2.2. How should you construct an outline?

The classical approach is to start with a blank piece of paper, and write down, in any order, all important ideas that occur to you concerning the paper. Ask yourself the obvious questions: "Why did I do this work?"; "What does it mean?"; "What hypothesis did I mean to test?"; "What ones did I actually test?"; "What were the results?"; "Did the work yield a new method or compound?"; "What measurements did I make?"; "What compounds?"; "How were they characterized?". Sketch possible equations, figures, and schemes. It is essential to try to get the major ideas written down. If you start the research to test one hypothesis, and decide, when you see what you have, that the data really seem to test some other hypothesis better, don't worry. Write them both down, and pick the best combinations of hypotheses, objectives and data. Often the objectives of a paper when it is finished are different from those used to justify starting the work. Much of good science is opportunistic and revisionist.

When you have written down what you can, start with another piece of paper and try to organize the jumble of the first one. Sort all of your ideas into three major heaps(1 - 3).

1. Introduction

Why did I do the work? What were the central motivations and hypotheses?

2. Results and Discussion

What were the results? How were compounds made and characterized? What was measured?

3. Conclusions

What does it all mean? What hypotheses were proved or disproved? What did I learn? Why does it make a difference?

Next, take each of these sections, and organize it on yet finer scale. Concentrate on organizing the *data*. Construct figures, tables, and schemes to present the data as clearly and compactly as possible. This process can be slow. I may sketch a figure 5 - 10 times in different ways, trying to decide how it is most clear (and looks best aesthetically).

Finally, put everything—outline of sections, tables, sketches of figures, equations—in good order.

When you are satisfied that you have included *all* the data (or that you know what additional data you intend to collect), and have a plausible organization, give the outline to me. Simply indicate where missing data will go, how you think (hypothesize) they will look, and how you will interpret them if your hypothesis is correct. I will take this outline, add my opinions, suggest changes, and return it to you. It usually takes 4 - 5 repeated attempts (often with additional experiments) to agree on an outline. When we *have* agreed, the data are usually in (or close to) final form (that is, the tables, figures, etc., in the outline will be the tables, figures, ⋯ in the paper).

You can then start writing, with some assurance that much of your prose will be used.

The key to efficient use of your and my time is that we start exchanging outlines and proposals as early in a project as possible. *Do not, under any circumstances, wait until the collection of data is "complete" before starting to write an outline.* No project is ever complete, and it saves enormous effort and much time to propose a plausible paper and outline as soon as you see the basic structure of a project. Even if we decide to do significant additional work before seriously organizing a paper, the effort of writing an outline will have helped to guide the research.

2.3. The outline

What should an outline contain?

1. *Title*
2. *Authors*
3. *Abstract*

Do *not* write an abstract. That can be done when the paper is complete.

4. *Introduction*

The first paragraph or two should be written out completely. Pay particular attention to the opening sentence. Ideally, it should state concisely the objective of the work, and indicate why this objective is important.

In general, the Introduction should have these elements:

- The *objectives* of the work.
- The justification for these objectives: Why is the work important?
- *Background*: Who else has done what? How? What have we done previously?
- *Guidance to the reader*. What should the reader watch for in the paper? What are the interesting high points? What strategy did we use?
- Summary/conclusion. What should the reader expect as conclusion? In advanced versions of the outline, you should also include all the sections that will go in the experimental section (at this point, just as paragraph subheadings).

5. *Results and Discussion*

The results and discussion are usually combined. This section should be organized according to major topics. The separate parts should have subheadings in boldface to make this organization clear, and to help the reader scan through the final text to find the parts that interest him or her. The following list includes examples of the phrases that might plausibly serve as section headings:

- Synthesis of Alkane Thiols
- Characterization of Monolayers
- Absolute Configuration of the Vicinal Diol Unit
- Hysteresis Correlates with Roughness of the Surface
- Dependence of the Rate Constant on Temperature
- The Rate of Self-Exchange Decreases with the Polarity of the Solvent

Try to make these section headings as specific and information-rich as possible. For example, the phrase "The Rate of Self-Exchange Decreases with The Polarity of

The Solvent" is obviously longer than "Measurement of Rates," but much more useful to the reader. In general, try to cover the major common points:

- Synthesis of starting materials
- Characterization of products
- Methods of characterization
- Methods of measurement
- Results (rate constants, contact angles, whatever)

In the outline, do not write any significant amount of text, but get all the data in their proper place: any text should simply indicate what will go in that section.

- Section Headings
- Figures (*with* captions)
- Schemes (with captions and footnotes)
- Equations
- Tables (correctly formatted)

Remember to think of a paper as a collection of experimental results, summarized as clearly and economically as possible in figures, tables, equations, and schemes. The text in the paper serves just to explain the data, and is secondary. The more information that can be compressed into tables, equations, etc., the shorter and more readable the paper will be.

6. Conclusion

In the outline, summarize the conclusions of the paper as a list of short phrases or sentences. Do not repeat what is in the Results section, unless special emphasis is needed. The Conclusions section should be just that, and not a summary. It should add a new, higher level of analysis, and should indicate explicitly the significance of the work.

7. Experimental

Include, in the correct order to correspond to the order in the Results section, all of the paragraph subheadings of the Experimental section.

2.4. In summary

- Start writing possible outlines for papers *early* in a project. Do not wait until the "end". The end may never come.
- Organize the outline and the paper around easily assimilated data-tables, equations, figures, schemes-rather than around text.

- Organize in order of importance, not in chronological order. An important detail in writing paper concerns the weight to be given to topics. Neophytes often organize a paper in terms of chronology: that is, they recount their experimental program, starting with their cherished initial failures and leading up to a climactic successful finale. *This approach is completely wrong. Start with the most important results*, and put the secondary results later, if at all. The reader usually does not care how you arrived at your big results, only what they are. Shorter papers are easier to read than longer ones.

3. Some Points of English Style

- Do not use nouns as adjectives:

Not:
ATP formation; reaction product
But:
formation of ATP; product of the reaction

- The word "this" must always be followed by a noun, so that its reference is explicit:

Not:
This is a fast reaction; This leads us to conclude
But:
This reaction is fast; This observation leads us to conclude

- Describe experimental results uniformly in the past tense:

Not:
Addition of water *gives* product.
But:
Addition of water *gave* product.

- Use the active voice whenever possible:

Not:
It was observed that the solution turned red.
But:
The solution turned red. *or*
We observed that the solution turned red.

- Complete all comparisons:

Not:
The yield was higher using bromine.
But:
The yield was higher using bromine than chlorine.

- Type all papers double-spaced (not single-or one-and-a-half spaced), and leave two space after colons, commas, and after periods at the end of sentences. Leave generous margins. (generally, 1.25″ on both sides & top & bottom).

Assume that we will write all papers using the style of the American Chemical Society. You can get a good idea of this style from three sources:

- *The journals*. Simply look at articles in the journals and copy the organization you see there.
- *Previous papers from the group*. By looking at previous papers, you can see exactly how a paper should "look". If what you wrote looks different, it probably is not what we want.
- *The ACS Handbook for Authors*. Useful, detailed, especially the section on references, pp. 173–229.

I also suggest you read Strunk and White, The Elements of Style (Macmillan: New York, 1979, 3rd ed.) to get a sense for usage. A number of other books on scientific writing are in the group library; these books all contain useful advice, but are not lively reading. There are also several excellent books on the design of graphs and figures.

Identification of CuO species in high surface area CuO—CeO$_2$ catalysts and their catalytic activities for CO oxidation

Meng-Fei Luo[*,†], **Yu-Peng Song**[†,‡], **Ji-Qing Lu**[†], **Xiang-Yu Wang**[‡] and **Zhi-Ying Pu**[†]

Zhejiang Key Laboratory for Reactive Chemistry on Solid Surfaces, Institute of Physical Chemistry, Zhejiang Normal University, Jinhua 321004, China, and Institute of Industrial Catalysis, Zhengzhou University, Zhengzhou 450052, China

Received: April 30, 2007; In Final Form: June 28, 2007

Nano-sized CuO—CeO$_2$ catalysts with high surface area (> 90 m$^2 \cdot$ g^{-1}) were prepared by a modified citrate sol—gel method with incorporation of N$_2$ thermal treatment. CO temperature-programmed reduction results indicated that there are three CuO species in the catalyst, namely the finely dispersed CuO, the bulk CuO and the Cu^{2+} in the CeO$_2$ lattice. Using CO oxidation as a model reaction, catalytic activity of each species was evaluated. It was found that the finely dispersed CuO species had the highest activity (183.3 mmol$_{CO} \cdot$ h$^{-1} \cdot$ g$_{Cu}^{-1}$), the bulk CuO had medium activity (100.4 mmol$_{CO} \cdot$ h$^{-1} \cdot$ g$_{Cu}^{-1}$), while the Cu^{2+} in the CeO$_2$ lattice had the lowest activity (21.3 mmol$_{CO} \cdot$ h$^{-1} \cdot$ g$_{Cu}^{-1}$). Furthermore, the Cu^{2+} species in the CeO$_2$ lattice could migrate to the catalyst surface to form finely dispersed CuO under high temperature calcination or under reaction conditions, which could consequently enhance the catalytic activity.

1. Introduction

In recent years, catalytic oxidation of CO has attracted considerable

---

* Corresponding author. Fax: +86 - 579 - 22×××××. E-mail address: mengfeiluo@zjnu.cn.
† Zhejiang Normal University.
‡ Zhengzhou University.

Identification of CuO Species in High Surface Area CuO−CeO$_2$ Catalysts and Their Catalytic Activities for CO Oxidation

Meng-Fei Luo,*,[†] Yu-Peng Song,[†,‡] Ji-Qing Lu,[†] Xiang-Yu Wang,[‡] and Zhi-Ying Pu[†]

Zhejiang Key Laboratory for Reactive Chemistry on Solid Surfaces, Institute of Physical Chemistry, Zhejiang Normal University, Jinhua 321004, China, and Institute of Industrial Catalysis, Zhengzhou University, Zhengzhou 450052, China

Received: April 30, 2007; In Final Form: June 28, 2007

Nanosized CuO−CeO$_2$ catalysts with high surface area (>90 m^2 g^{-1}) were prepared by a modified citrate sol−gel method with incorporation of N$_2$ thermal treatment. CO temperature-programmed reduction results indicated that there are three CuO species in the catalyst, namely, the finely dispersed CuO, the bulk CuO, and the Cu^{2+} in the CeO$_2$ lattice. Using CO oxidation as a model reaction, catalytic activity of each species was evaluated. It was found that the finely dispersed CuO species had the highest activity (183.3 mmol$_{CO}$ g$_{Cu}^{-1}$ h^{-1}) and the bulk CuO had medium activity (100.4 mmol$_{CO}$ g$_{Cu}^{-1}$ h^{-1}), while the Cu^{2+} in the CeO$_2$ lattice had the lowest activity (21.3 mmol$_{CO}$ g$_{Cu}^{-1}$ h^{-1}). Furthermore, the Cu^{2+} species in the CeO$_2$ lattice could migrate to the catalyst surface to form finely dispersed CuO under high-temperature calcination or under reaction conditions, which could consequently enhance the catalytic activity.

1. Introduction

In recent years, catalytic oxidation of CO has attracted considerable attention due to many applications such as pollution control for vehicle exhaust, trace CO removal in the enclosed atmospheres, gas purification for CO$_2$ lasers, CO gas sensors, and fuel cell.[1,2] Noble metal catalysts such as Au, Pt, and Pd have been proved very effective for CO oxidation at low temperature.[1,3−5] However, due to the high cost of noble metals and sensitivity to sulfur poisoning, more and more research is focusing on new catalysts containing cheap transition metals. Among them, copper catalyst was found to be an excellent base metal catalyst for CO oxidation.[6,7] CuO−CeO$_2$ catalysts have been widely studied for various reactions such as NO reduction, complete CO oxidation, preferential oxidation (PROX), the water-gas shift (WGS), and the wet oxidation of phenol due to high activity and selectivity for these reactions.[8−11] Avgouropoulos et al. reported that CuO−CeO$_2$ catalyst had the same activity as Pt/Al$_2$O$_3$ catalyst.[12] Sedmak et al. prepared nanostructured Cu$_{0.1}$Ce$_{0.9}$O$_{2-y}$ by a sol−gel method, which exhibited excellent activity and selectivity for preferential oxidation of CO in excess H$_2$ at low temperature.[8] Due to the promising advantage of low price and high activity, CuO−CeO$_2$ catalysts are expected to substitute for the noble metal catalysts in the future.[13]

For CO oxidation, there are two facts that govern the activity of the CuO−CeO$_2$ catalysts, namely, surface area and active sites of the catalyst. Although much attention is paid to the CuO−CeO$_2$ catalysts, CuO−CeO$_2$ catalysts with high surface area (≥60 m^2 g^{-1}) were rarely reported. It is well-known that a catalyst with higher surface area generally gives better catalytic activity, due to the fact that it can provide more active sites.[14,15] Our previous work has reported high surface area nanosized CuO−CeO$_2$ catalysts prepared by a surfactant-templated method.[16] The catalysts show high catalytic activity for low-temperature CO oxidation and selective oxidation of CO in excess H$_2$. Another work of our group reported a modified sol−gel method to produce Ce$_{0.8}$Pr$_{0.2}$O$_y$ solid solution with ultrafine crystalline sizes and high specific surface area.[17] Compared to the surfactant-templated method, the modified sol−gel method has advantages such as simple preparation and high yield of catalyst.

It was believed that CuO species are the active sites for CO oxidation;[18] however, the existing state of CuO in the CuO−CeO$_2$ catalyst is usually complex, and contribution of each CuO species to catalytic activity should be different. Our previous work found that in the CuO−CeO$_2$ catalysts prepared by surfactant-templated method the highly dispersed CuO was the active phase for CO oxidation.[16] However, contribution of other CuO species to the catalytic performance remains unsolved.

To deepen the investigation, in this work, the CuO−CeO$_2$ catalysts with high surface area were prepared using a modified sol−gel method. These catalysts show high activities for CO oxidation. The CuO species in the catalyst were distinguished, and their contributions to the catalytic activity were evaluated. Moreover, translation of these CuO species during reaction was also studied.

2. Experimental Section

2.1. Catalyst Preparation. The catalysts were prepared using two different methods: the conventional citrate sol−gel method and the modified citrate sol−gel method, respectively.

CuO−CeO$_2$ catalysts with different CuO contents (5, 10, 20, and 50 mol %) were prepared using the modified citrate sol−gel method, as described elsewhere.[19] A mixture of Ce(NO$_3$)$_3$·6H$_2$O and Cu(NO$_3$)$_2$·3H$_2$O with different molar ratio of Cu/(Ce + Cu) was dissolved into deionized water. Then citric acid was added to the premixed cerium and copper nitrate solution while stirring, with a molar ratio of acid/(Ce + Cu) = 2. After that, the solution was heated in a water bath until a viscous gel

---

* Corresponding author. Fax: +86-579-22×××××. E-mail address: mengfeiluo@×××.cn.
[†] Zhejiang Normal Univer×××.
[‡] Zhengzhou Univer×××.

attention due to many applications such as pollution control for vehicle exhaust, trace CO removal in the enclosed atmospheres, gas purification for CO_2 lasers, CO gas sensors and fuel cell[1,2]. Noble metal catalysts such as Au, Pt and Pd have been proved very effective for CO oxidation at low temperature[1,3-5]. However, due to the high cost of noble metals and sensitivity to sulfur poisoning, more and more researches are focusing on new catalysts containing cheap transition metals. Among them, copper catalyst was found to be an excellent base metal catalyst for CO oxidation[6,7]. $CuO-CeO_2$ catalysts have been widely studied for various reactions such as NO reduction, complete CO oxidation, preferential oxidation (PROX), the water-gas-shift (WGS) and the wet oxidation of phenol due to high activity and selectivity for these reactions[8-11]. Avgourpoulos et al. reported that $CuO-CeO_2$ catalyst had same activity as Pt/Al_2O_3 catalyst[12]. Sedmak et al. prepared nanostructured $Cu_{0.1}Ce_{0.9}O_{2-y}$ by a sol-gel method, which exhibited excellent activity and selectivity for preferential oxidation of CO in excess H_2 at low temperature[8]. Due to the promising advantage of low price and high activity, $CuO-CeO_2$ catalysts are expected to substitute for the noble metal catalysts in the future[13].

For CO oxidation, there are two facts that govern the activity of the $CuO-CeO_2$ catalysts, namely surface area and active sites of the catalyst. Although much attention is paid on the $CuO-CeO_2$ catalysts, $CuO-CeO_2$ catalysts with high surface area ($\geqslant 60 \text{ m}^2 \cdot \text{g}^{-1}$) were rarely reported. It is well known that catalyst with higher surface area generally gives better catalytic activity, due to the fact that it can provide more active sites[14,15]. Our previous work has reported high surface area nanosized $CuO-CeO_2$ catalysts prepared by a surfactant-templated method[16]. The catalysts show high catalytic activity for low temperature CO oxidation and selective oxidation of CO in excess H_2. Another work of our group reported a modified sol-gel method to produce $Ce_{0.8}Pr_{0.2}O_y$ solid solution with ultrafine crystalline sizes and high specific surface area[17]. Compared to the surfactant templated method, the modified sol gel method has advantages such as simple preparation and high yield of catalyst.

It was believed that CuO species are the active sites for CO oxidation[18], however, the existing state of CuO in the $CuO-CeO_2$ catalyst is usually complex, and contribution of each CuO species to catalytic activity should be different. Our previous work found that in the $CuO-CeO_2$ catalysts prepared by surfactant templated method the highly dispersed CuO was the active phase for CO oxidation[16]. However, contribution of other CuO species to the catalytic performance remains unsolved.

In order to deepen the investigation, in this work, the CuO—CeO_2 catalysts with high surface area were prepared using a modified sol-gel method. These catalysts show high activities for CO oxidation. The CuO species in the catalyst were distinguished and their contributions to the catalytic activity were evaluated. Moreover, translation of these CuO species during reaction was also studied.

2. Experimental

2.1. Catalyst Preparation

The catalysts were prepared using two different methods: the conventional citrate sol-gel method and the modified citrate sol-gel method, respectively.

CuO—CeO_2 catalysts with different CuO contents (5 mol %, 10 mol %, 20 mol %, 50 mol %) were prepared using the modified citrate sol-gel method, as described elsewhere[19]. A mixture of $Ce(NO_3)_3 \cdot 6H_2O$ and $Cu(NO_3)_2 \cdot 3H_2O$ with different molar ratio of Cu/(Ce + Cu) was dissolved into deionized water. Then citric acid was added to the premixed cerium and copper nitrate solution while stirring, with a molar ratio of acid/(Ce + Cu) = 2. After that, the solution was heated in water bath until a viscous gel was obtained. In this process, the mixture color turned from blue to green. The gel was dried at 110 ℃ overnight to form a spongy material, i.e. catalyst precursor. Then the precursor was put in a tube furnace and heated in N_2 at 800 ℃ for 2 h to result in a black mixture of CuO—CeO_2 and carbon powders (the intermediate mixture). Subsequently, the intermediate mixture was calcined in air at 400 ℃ for 4 h to burn the carbon species. The actual CuO content in the catalyst was analyzed by atomic absorption spectrophotometry (AAS). In designation, a N8A4-5 catalyst means a CuO—CeO catalyst treated in N_2 at 800 ℃ and in air at 400 ℃, and the CuO content in the catalyst is 5 mol %.

For comparison, a series of CuO—CeO_2 catalysts with 10 mol % CuO content were prepared according to the conventional sol-gel method[20]. The precursor was directly calcined in air at 400 ℃ or 800 ℃, denoted as A4-10 or A8-10, respectively.

For samples treated with nitric acid, 1 g of catalyst (N8A4-10) was immersed in concentrated nitric acid (15 mL of 50 % HNO_3/g catalyst) for 2 h and then filtered, washed with plenty of distilled water to remove the residual nitric acid or other impurities. It was dried at 100 ℃, 200 ℃ or calcined at

600 ℃, denoted as N8A4-10H1, N8A4-10H2 and N8A4-10H6, respectively. Meanwhile, sample A8-10 was treated with nitric acid in the same manner and was dried at 100 ℃, denoted as A8-10H1.

2.2. Catalyst Characterization

X-ray diffraction (XRD) patterns were collected on a PHILIPS PW 3040/60 powder diffractometer using Cu Kα radiation. The working voltage of the instrument was 40 kV and the current was 40 mA. The intensity data were collected at 25 ℃ in a 2θ range from 20° to 130° with a scan rate of 1.2° · min^{-1}. The microstructural parameters of samples were determined by the Rietveld method[21] using MAUD software (Material Analysis Using Diffraction)[22]. CeO_2 annealed at 1450 ℃ for 10 h in air was used as a non-intrinsic broadening sample to measure the instrument function and to extract the micro-structure values of the catalyst.

Specific surface areas (S_{BET}) of the catalysts were calculated from a multipoint Braunauer Emmett Teller (BET) analysis of the nitrogen adsorption and desorption isotherms at 77 K recorded on an Autosorb-1 apparatus.

Transmission electron microscopy (TEM) investigations were carried out using a JEM-1200 EX microscope operated at 80 kV.

X-ray photon spectroscopy (XPS) experiments were carried out on a RBD upgraded PHI-5000C ESCA system (Perkin-Elmer) with Mg Kα radiation ($h\nu$ = 1253.6 eV). The pass energy was fixed at 46.95 eV to ensure sufficient sensitivity. Binding energies were calibrated by using the containment carbon (C 1s, 284.6 eV). The data analysis was performed using the RBD AugerScan 3.21 software (RBD Enterprises).

Reducibility of CuO—CeO_2 catalysts were measured by CO temperature-programmed reduction (CO—TPR). 50 mg of sample was placed in a quartz reactor which was connected to a homemade TPR apparatus and the reactor was heated from 30 ℃ to 600 ℃ with a heating rate of 20 ℃ min^{-1}. The reaction mixture consists of 5% CO and 95% Ar. The amount of CO consumption and the signal of CO_2 were monitored by a Balzers Omnistar 200 mass spectrometer at $m/e = 44$.

2.3. Catalytic Activity Measurement

The catalytic activity of CO oxidation was evaluated in a fixed bed reactor (6 mm i.d.) using 250 mg of catalyst (20 – 40 mesh). The feed gas consists of

1% CO and 1% O_2 in N_2 with a total flow rate of 40 mL (NTP) min^{-1}, corresponding to a space velocity (S.V.) of 9600 mL $\cdot$ g^{-1} $\cdot$ h^{-1}. The catalysts were directly exposed to reaction gas as the reactor temperature was stabilized at the reaction temperature without any pretreatment. The reaction temperature was monitored by a thermocouple placed in the middle of the catalyst bed. The CO concentration in the reactor effluent was analyzed using an Agilent 6850 gas chromatograph equipped with a TCD detector attached to an HP PLOT column (30 m × 0.32 mm × 12 μm).

3. Results

3.1. Structural Characterization

The specific surface areas of $CuO-CeO_2$ catalysts with 10 mol % Cu content prepared by two methods are summarized in Table 1. The surface area of sample N8A4-10 is 131 $m^2 \cdot g^{-1}$, which is much larger than those of samples A4-10 (60 $m^2 \cdot g^{-1}$) and A8-10 (10 $m^2 \cdot g^{-1}$). The dramatic decline in surface area for the A8-10 indicates that increasing calcination temperature leads to particle sintering during calcination at high temperature [23]. By comparing surface areas of the three catalysts, it can be concluded that the modified sol-gel method with the incorporation of N_2 thermal treatment can afford high surface area carriers.

The surface areas for the $CuO-CeO_2$ catalysts with various Cu content prepared by the modified sol-gel method are also listed in Table 1. It can be seen that the surface area increases with increasing CuO content when CuO content is below 10 mol %, which implies that the introduction of Cu enhances the surface area. However, the surface area decreases with CuO content increasing from 10 mol % to 50 mol %. It is likely due to the formed bulk CuO when CuO content is over 10 mol %.

Table 1 BET surface area, cell parameters and crystallite size of $CuO-CeO_2$ catalysts

Samples	Phase composition	S_{BET} ($m^2 \cdot g^{-1}$)	Cell parameter of CeO_2 (nm)	Crystal. size of CeO_2 (nm)		
				$d(111)$	$d(200)$	$d(220)$
CeO_2	CeO_2	73	0.5423(1)	10.84	10.86	10.85
N8A4-5	CeO_2	92	0.5421(1)	6.94	7.42	7.06
N8A4-10	CeO_2	131	0.5418(2)	7.08	7.53	7.19

(Continued)

Samples	Phase composition	S_{BET} ($m^2 \cdot g^{-1}$)	Cell parameter of CeO_2 (nm)	Crystal. size of CeO_2 (nm)		
				$d(111)$	$d(200)$	$d(220)$
N8A4-20	CeO_2 + CuO	95	0.5415(2)	6.52	6.59	6.53
N8A4-50	CeO_2 + CuO	90	0.5415(2)	5.93	5.90	5.92
A4-10	CeO_2	60	0.5419(1)	7.99	8.29	8.20
A8-10	CeO_2 + CuO	10	0.5418(1)	63.63	63.58	63.71

Figure 1 shows XRD patterns of the CuO—CeO_2 catalysts. The patterns of catalysts with 10 mol % Cu content are presented in Figure 1a, where typical diffraction peaks of CeO_2 are observed in all the samples. For sample A8-10, two weak peaks characteristic of CuO phase are observed at 35.6° and 38.8°, indicating that bulk CuO is formed in the catalyst. Figure 1b presents XRD patterns of the CuO—CeO_2 catalysts with various Cu content prepared by the modified method.

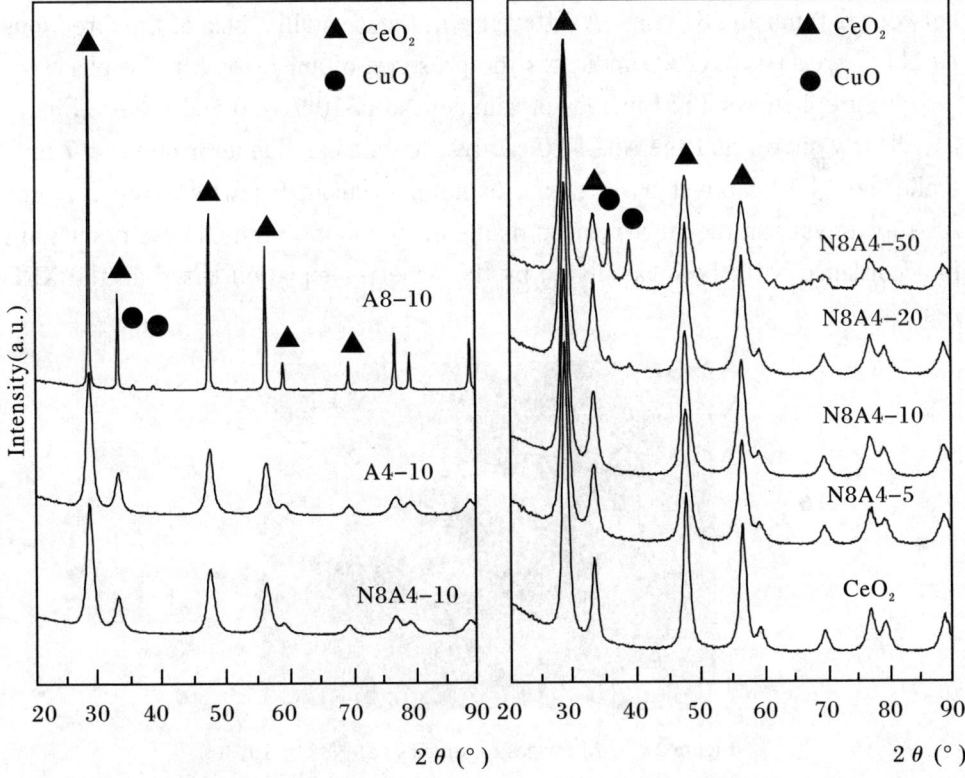

Figure 1 XRD patterns of CuO—CeO_2 catalysts: (a) catalysts with 10 mol % Cu content; (b) catalysts with various Cu content.

CeO_2 phase is observed in all the samples. However, diffraction peaks due to CuO are not detected with CuO content lower than 20 mol %. When the CuO content increases up to 20 mol %, weak diffraction peaks ascribed to bulk CuO appears, and the intensities of these peaks increase apparently with increasing CuO content up to 50 mol %.

Phase composition, cell parameter and crystallite size measurements of the $CuO-CeO_2$ catalysts determined by XRD are also listed in Table 1. It can be seen that all the samples contain a cubic CeO_2 phase. In addition, for high CuO content samples such as N8A4-20 and N8A4-50 or the sample calcined at high temperature (A8-10), a monoclinic CuO phase is also present. It is clear that cell parameter decreases with increasing CuO content up to 20 mol %, but remains almost constant between 20 mol % and 50 mol %. The crystallite sizes of N8A4-10, A4-10 and A8-10 are about 7.3 nm, 8.2 nm and 63.6 nm, respectively. For the samples with different CuO contents, the crystallite size of sample CeO_2 is the largest (10.9 nm), while those of other samples are between 6.0 nm and 8.0 nm. A difference in the crystallite size of the directions $d(111)$, $d(200)$, $d(220)$ indicates the presence of anisotropy in the phase.

Figure 2 shows TEM images of samples N8A4-10, A4-10 and A8-10. Figure 2a. clearly shows that the N8A4-10 catalyst has a mean diameter of about 7 nm, while the A4-10 catalyst has a mean diameter of about 9 nm. In contrast, the A8-10 catalyst has the largest mean diameter of about 60 nm. These results are in accordance with those calculated by the Scherrer equation based on the XRD results.

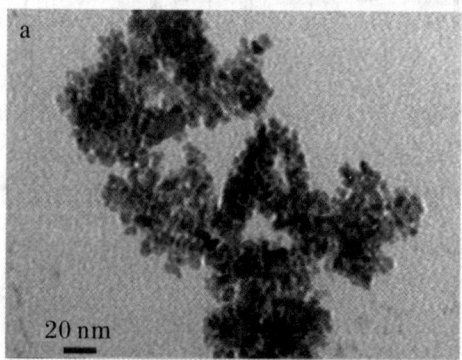

Figure 2 TEM images of samples (a) N8A4 − 10, (b) A4 − 10 and (c) A8 − 10.

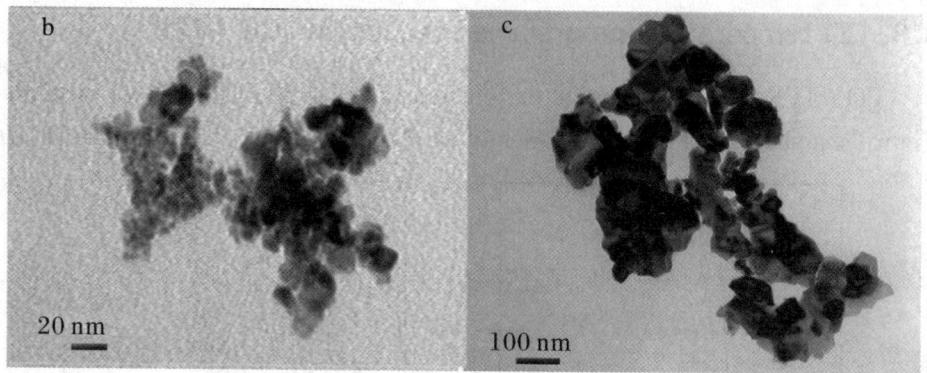

Continued Figure 2

3.2. Catalytic Testing

Figure 3 shows light-off curves of CO oxidation over the CuO—CeO_2 catalysts prepared by different methods. The catalytic activities of catalysts with 10 mol% Cu content are presented in Figure 3a. The T_{90} (the temperature when the conversion is 90%) for the samples N8A4-10, A4-10 and A8-10 are 100 ℃, 110 ℃ and 120 ℃, respectively. The catalytic activities for CO oxidation of series CuO—CeO_2 catalysts prepared by the modified method are shown in Figure 3b. The activities of pure CuO and CeO_2 are quite low. With the incorporation of CuO, the activity of CuO—CeO_2 catalysts is improved significantly. The catalytic activity increases with the CuO content increasing until CuO content is 10 mol %, however, catalysts with higher CuO content give almost constant activities.

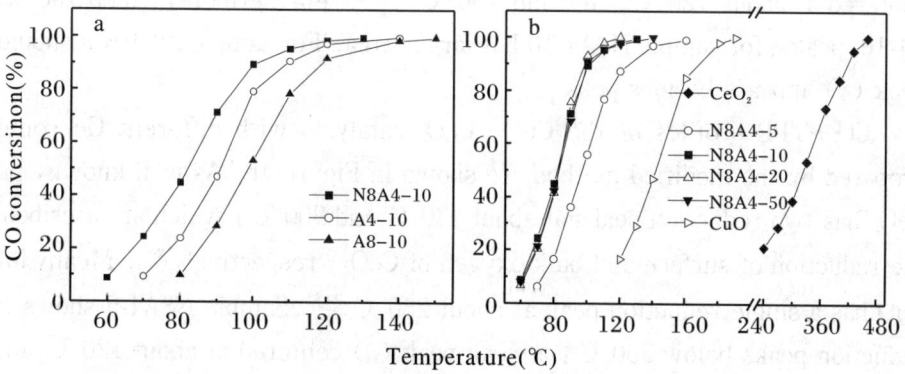

Figure 3 Catalytic activities of CuO—CeO_2 catalysts for CO oxidation: (a) catalysts with 10 mol % Cu content; (b) catalysts with various Cu content.

3.3. CO temperature-programmed reduction (CO—TPR)

CO—TPR is used to characterize reducibility of the CuO—CeO$_2$ catalysts. Figure 4 a shows CO—TPR profiles of samples N8A4-10, A4-10 and A8-10.

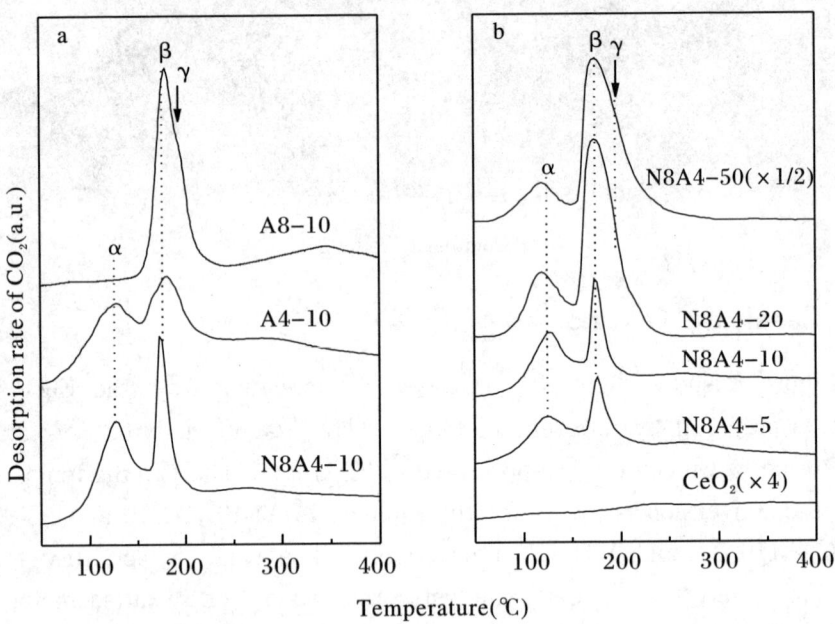

Figure 4 CO—TPR profiles of CuO—CeO$_2$ catalysts. (a) catalysts with 10 mol % Cu content; (b) catalysts with various Cu content.

From the figure, it can be seen that two low temperature (below 200 ℃) reduction peaks (α and β) appear in both profiles of N8A4-10 and A4-10, centered at about 120 ℃ (α) and 180 ℃ (β). Furthermore, compared with A4-10, peak α for sample N8A4-10 has larger area. For sample A8-10, a shoulder peak (γ) appears besides peak β.

CO—TPR profiles of the CuO—CeO$_2$ catalysts with different Cu content prepared by the modified method are shown in Figure 4b. As well known, pure CeO$_2$ has two reduction peaks at about 430 ℃ and 900 ℃, which are ascribed to the reduction of surface and bulk oxygen of CeO$_2$, respectively[20]. Meanwhile, CuO has a single reduction peak at about 280 ℃ [24]. Sample N8A4-5 shows two reduction peaks below 200 ℃: a weak peak (α) centered at about 120 ℃ and a strong peak (β) centered at about 175 ℃. With CuO content increasing from 5 mol % to 10 mol %, the intensities of the peak α and β increase, however, when the CuO content ranges from 10 mol % to 50 mol %, peak α does not change significantly and peak β becomes broader, together with a shoulder peak (γ) appeared at about 195 ℃.

4 Discussion

4.1. Identification of Cu Species in the CuO—CeO$_2$ Catalysts

As mentioned earlier (Figure 1), no bulk CuO phase is observed at lower CuO content. It is likely because the solid solution is formed [11] or the CuO particles are finely dispersed on the surface of CeO$_2$, which are too small to be detected by X-ray diffraction method [24]. Comparatively, the presence of bulk CuO phase when CuO content increases to 20 mol % should be ascribed to the aggregation of CuO species on the surface. It is also found that crystallite sizes of the catalysts prepared by the modified sol-gel method are smaller than those prepared by the conventional sol-gel method, confirmed by the XRD (Table 1) and TEM (Figure 2) results. It is due to the fact that with the modified method the precursors are firstly treated in N$_2$ at high temperature and the organic citrate does not burn up but decompose into ultrafine carbon powders, which in turn can encapsulate or isolate Ce—Cu oxides and prevent them from sintering. Moreover, the incorporation of CuO inhibits the crystal growth of CeO$_2$, but the CeO$_2$ crystal have no apparent growth with increasing CuO content, which is in agreement with the result by Fu et al[5]. It is also found that the crystal size of CeO$_2$ decreases with Cu content, which was also reported in previous literature; 26 however, a proper interpretation of the change in particle size with CuO content remains unsolved. The cell parameter decreases with the increase of CuO content, this is because the Cu^{2+} ionic radius (0.072 nm) is smaller than that of Ce^{4+} (0.097 nm), when CuO partially incorporates into the CeO$_2$ lattice and Cu^{2+} replaces Ce^{4+}, there is a reduction in the cell parameter of CeO$_2$. This indicated that the Cu$_x$Ce$_{1-x}$O$_{2-\delta}$ solid solution is formed in these catalysts[27].

CO—TPR results (Figure 4) reveal that two reduction peaks (α and β) were observed in all the profiles of CuO—CeO$_2$ catalysts. Besides, another reduction peak γ appears in the profiles of A8-10, N8A4-20 and N8A4-50. There are different ascriptions to the low-temperature reduction peaks. Avgouropoulos and Ioannides regards the low temperature peak as reduction of copper species[26] strongly interacting with CeO$_2$ and higher temperature peak as reduction of less or non-interacting CuO species. Zou et al. attributes the anterior peak to the reduction of cluster copper species and the latter peak to the reduction of isolated Cu^{2+} ions[27]. Two reduction peaks have been reported by Luo et al.[18], where peak α is ascribed to finely dispersed CuO, while peak β is assigned to the larger particles of the bulk CuO. Same results are also reported by Shiau et al.[28,29]. In this work, the XRD results reveal that the formation of $Cu_xCe_{1-x}O_{2-\delta}$ solid solution in all the samples and bulk CuO in high Cu-content samples. Also, it is known that the finely dispersed CuO is easy to be reduced.

Thus peak is due to the reduction of finely dispersed CuO, peak β is due to the reduction of the Cu^{2+} in $Cu_xCe_{1-x}O_{2-\delta}$ solid solution. Peak γ is attributed to bulk CuO because the XRD results clearly show that bulk CuO phase appears for the samples A8-10, N8A4-20 and N8A4-50. Therefore, it can be concluded that there are three CuO species in the CuO-CeO_2 catalysts: the finely dispersed CuO species on the surface of CeO_2, the Cu^{2+} in CeO_2 lattice and the bulk CuO phase, similar result was also reported in our previous work[16].

4.2. Contributions of different CuO species to catalytic activity

The CuO—CeO_2 catalysts reported in this work have slightly higher catalytic activities than other CuO—CeO_2 catalysts for low temperature CO oxidation, indicating that the modified citrate sol-gel method can provide catalyst with high catalytic activity[15,33]. In our previous work, it was reported that the finely dispersed CuO is the active site for CO oxidation[16,18], which is also confirmed in this work. Seen from the activities of the catalysts for CO oxidation (Figure 3), sample N8A4-10 has better activity than A4-10 and A8-10, while samples N8A4-10, N8A4-20 and N8A4-50 have similar activities. The CO—TPR results (Figure 4) indicate that sample N8A4-10 has larger peak area of α than A4-10 and A8-10, while N8A4-10, N8A4-20 and N8A4-50 have similar peak area of α, suggesting that the catalytic activity is mainly related to the finely dispersed CuO species. However, sample A8-10 without finely dispersed CuO species still shows some activity for CO oxidation although it is low, implying that other CuO species may also play a role in the reaction.

To clarify contributions of the three CuO species to CO oxidation, the nitric acid treatment on the N8A4-10 and A8-10 was conducted. It is well known that nitric acid can dissolve the free CuO[18]. The CuO content of samples N8A4-10H1 and A8-10H1 is 4.95 and 4.96 mol% by AAS analysis, respectively. CO oxidation activities of samples N8A4-10 and A8-10 before and after nitric acid treatment are presented in Figure 5. It can be seen that after the acid treatment, the activities of the samples N8A4-10H1 and A8-10H1 declined dramatically compared to the untreated catalysts. However, it should be noticed that these two acid treated samples have almost same activity. In order to explain the results, CO—TPR experiments of these samples were carried out and the profiles are shown in Figure 6. As mentioned above, two reduction peaks α and β appear in the profile of sample N8A4-10, while after the acid treatment, peak α disappears and only peak β remains. Compared to the A8-10, the peak γ is absent and the peak β becomes more symmetry in the profile of the A8H1-10. Therefore, the activity results from the Cu^{2+} in the CeO_2 lattice. Also, the fact that A8H1-10 and N8A4H1-10 have same peak area of β could explain why these two catalysts have same activity. These results again confirm that the peak α is the reduction of finely dispersed CuO species and the peak γ is the reduction of the bulk CuO, while the peak β is due to the reduction of Cu^{2+} in the CeO_2 lattice.

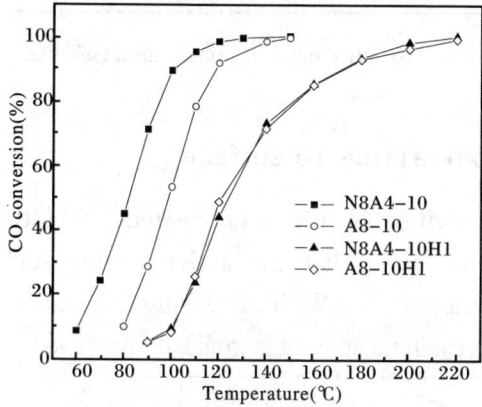

Figure 5 Catalytic activities of samples N8A4-10 and A4-10 before and after nitric acid treatment.

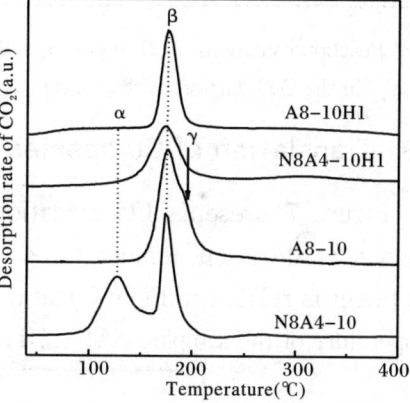

Figure 6 CO—TPR profiles of samples N8A4-10 and A4-10 before and after nitric acid treatment.

Specific reaction rates of the CuO—CeO_2 catalysts normalized by Cu content for CO oxidation at 100 ℃ were calculated and the results are shown in Table 2. With different CO conversions at 100 ℃, these samples show different specific rates. The N8A4-10 shows the highest rate (101.8 $mmol_{CO} \cdot g_{Cu}^{-1} \cdot h^{-1}$), the A8-10 shows medium rate (60.6 $mmol_{CO} \cdot g_{Cu}^{-1} \cdot h^{-1}$), while the N8A4-10H1 and A8-10H1 show the lowest rates (21.3 $mmol_{CO} \cdot g_{Cu}^{-1} \cdot h^{-1}$). Considering that the N8A4-10 has a phase mixture of finely dispersed CuO and Cu^{2+} in the CeO_2 lattice, the A8-10 has a mixture of Cu^{2+} in the CeO_2 lattice and bulk CuO, while the N8A4-10H1 or A8-10H1 has Cu^{2+} in the CeO_2 lattice, the overall rates for each catalyst was calculated and listed in Table 2. Based on these results, the specific rate of the individual CuO species could be calculated. It is clear that the finely dispersed CuO has the largest contribution to the activity (183.3 mmol$_{CO}$ · g$_{Cu}^{-1}$ · h^{-1}), bulk CuO has medium one (100.4 mmol$_{CO}$ · g$_{Cu}^{-1}$ · h^{-1}) while the Cu^{2+} in the CeO$_2$ lattice has the least (21.3 mmol$_{CO}$ · g$_{Cu}^{-1}$ · h^{-1}).

Table 2 Specific rates of different catalysts for CO oxidation at 100 ℃ ^{a}

Sample	Cu content (wt %)			CO conv. (%)	Overall rate (mol$_{CO}$ · g$_{Cu}^{-1}$ · h^{-1})	Specific rate (mol$_{CO}$ · g$_{Cu}^{-1}$ · h^{-1})		
	a+b	b	b+c			a	b	c
N8A4-10	3.76	-	-	89.3	101.8	183.3	21.3	-
A8-10	-	-	3.76	53.2	60.6	-	21.3	100.4
N8A4-10H1	-	1.89	-	9.4	21.3	-	21.3	-
A8-10H1	-	1.89	-	9.4	21.3	-	21.3	-

^{a} Reaction condition: 0.25 g cat, gas flow rate: 40 mL · min^{-1}; a: finely dispersed CuO; b: Cu^{2+} in the CeO$_2$ lattice; c: bulk CuO

4.3. Translation of Cu species from lattice to surface

Figure 7 presents CO oxidation activities of the acid treated N8A4-10 calcined at different temperatures. It is found that the activity after acid treatment is related to the calcination temperature. With increasing calcination temperature of the sample N8A4-10H1, the catalytic activity is gradually improved.

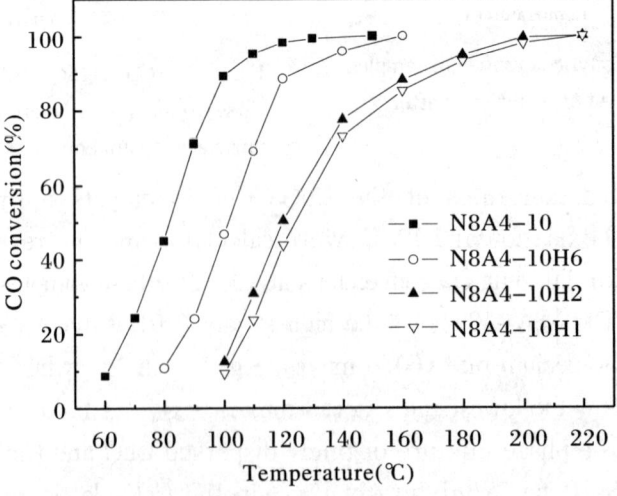

Figure 7 Catalytic activities of sample N8A4-10 before and after nitric acid treatment.

The enhancement in catalytic activity could be explained by CO—TPR profiles of these samples presented in Figure 8. It is obvious that after the acid treatment, the peak α of samples N8A4-10H1 and N8A4-10H2 was absent, while the peak β kept constant. It is interesting that for the acid treated sample calcined at 600 ℃ (N8A4-10H6), peak α corresponding to the finely dispersed CuO species appears again, accompanied with the decrease of peak β which corresponds to the Cu^{2+} in the CeO_2 lattice. However, the overall peak area has not changed, indicating that the reducible CuO species content was unchanged. Therefore, the presence of the finely dispersed CuO in the N8A4-10H6 must be related to the Cu^{2+} in the lattice. That is to say, when the acid treated sample was calcined at 600 ℃, partial Cu^{2+} in the $Cu_x Ce_{1-x} O_{2-\delta}$ solid solution migrated to the surface to form finely dispersed CuO.

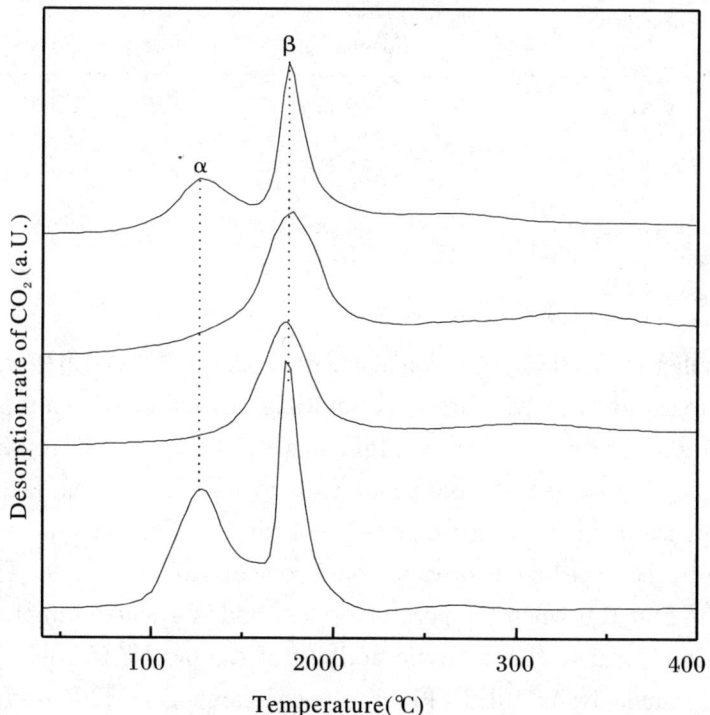

Figure 8 CO—TPR profiles of sample N8A4-10 before and after nitric acid treatment.

To better understand the change of surface composition of the catalyst caused by the acid treatment, XPS analysis was conducted. Table 3 summarized the surface concentrations of Cu before and after the acid treatment. It is very clear that for the untreated sample (N8A4-10), the surface Cu concentration (15.26 mol%) is much higher than the nominal one (9.58 mol%), due to the enrichment of the CuO species on the surface, while after acid treatment, the surface Cu concentration of sample N8A4-10H1 (5.73 mol%) is close to the nominal one (4.95 mol%), indicating that the surface CuO species are completely removed. Meanwhile, the surface Cu concentration of sample N8A4-10H6 is 11.17 mol%, which is much higher than the nominal one (4.95 mol%). These results indicate the enrichment of CuO species on the surface, which is in good agreement with the CO—TPR results, and strongly suggest that the surface CuO species plays dominant role in CO oxidation and the change in activity is due to the enrichment of the finely dispersed CuO on the surface.

Table 3 Comparison of Cu content before and after nitric acid treatment

Samples	Cu/(Ce+Cu) (mol %)	
	Nominal (AAS)	XPS
N8A4-10	9.58	15.26
N8A4-10H1	4.95	5.73
N8A4-10H2	4.95	5.99
N8A4-10H6	4.96	11.17

To further prove the conclusion above, the sample N8A4-10H1 was used for CO cyclic oxidation. In this process, multiple conversion-temperatures scans were operated, the reaction temperature increased from 80 ℃ to 220 ℃ with the CO conversion increased from about 10% to 100% and then decreased in a similar stepwise manner, which denoted as a run[31]. The catalytic activities of the sample in the experiment process could be seen from Figure 9. The activity is low in the first run when temperature rises, and is gradually improved in the process. After 4 runs, the catalytic activity of sample N8A4-10H1 is close to that of the sample N8A4-10H6. Figure 10 compares CO—TPR profiles of the N8A4-10H6 and the sample N8A4-10H1 after 4 runs. It is found that after 4 runs, a reduction peak (α) at 130 ℃ appears, corresponding to the finely dispersed CuO species. These results again suggest that the Cu^{2+} species migrate from the CeO_2 lattice to the surface during the reaction.

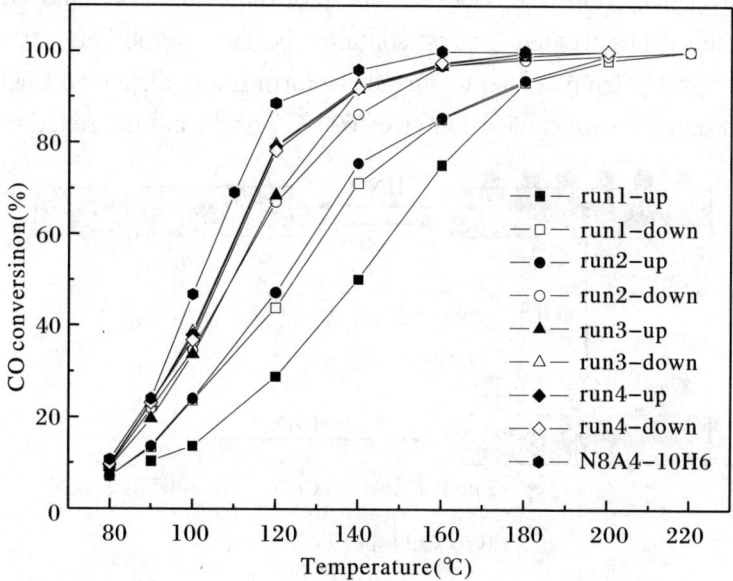

Figure 9 Cyclic activities of N8A4-10H1 for CO oxidation.

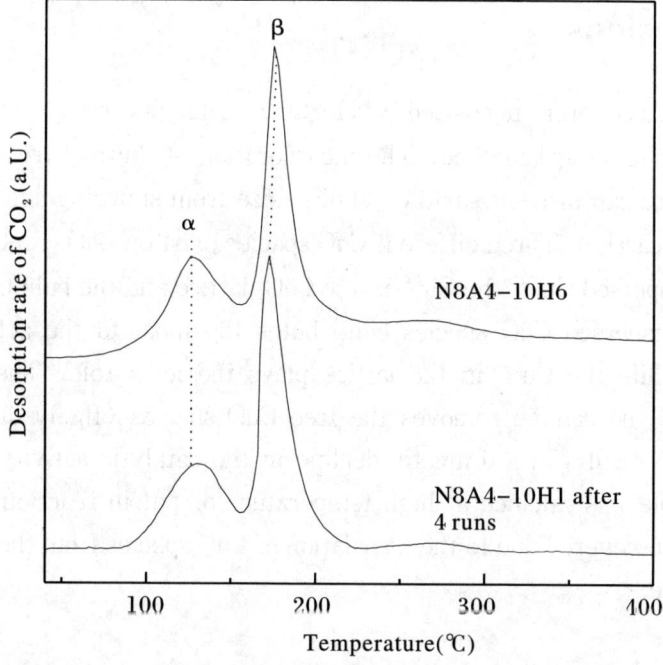

Figure 10 CO—TPR profiles of the N8A4-10H6 and N8A4-10H1 after 4 runs.

Scheme 1 illustrates the shift between the three CuO species in the catalyst. The finely dispersed CuO could be translated to the bulk CuO at high temperature calcination (i.e. 800 ℃). Both of them could be removed by nitric acid treatment. The treated catalyst contains the Cu^{2+} in the CeO_2 lattice, while part of it migrates from lattice to surface to form finely dispersed CuO particles under high temperature calcination (i.e. 600 ℃) or during the reaction process.

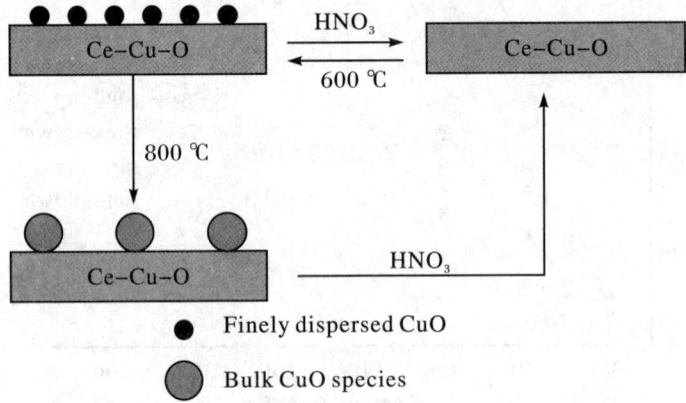

Scheme 1　Shift between the three CuO species in the catalyst.

5. Conclusions

High-surface area nanosized $CuO—CeO_2$ catalysts were obtained by a modified citrate sol-gel method. The incorporation of thermal treatment under N_2 atmosphere can prevent small crystallite size from sintering and offer a high surface area carrier. Three different CuO species exist on $CuO—CeO_2$ catalysts: the finely dispersed CuO, the Cu^{2+} in the CeO_2 lattice and the bulk CuO species. The finely dispersed CuO species contributes the most to the activity of CO oxidation, while the Cu^{2+} in the lattice plays the least role. The nitric acid treatment on the sample removes the free CuO species (finely dispersed and bulk), which results in a dramatic decline in the catalytic activity. When the treated sample was calcined at high temperature or put in reaction cycles, the activity was recovered due to the translation of CuO species from the CeO_2 lattice to the surface.

Acknowledgment

This work is financially supported by the Natural Science Foundation of China (Grant 20473075) and the Zhejiang Provincial Nature Science Foundation of China (Grant Z404383).

References and Notes

[1] Zhu, H. Q.; Qin, Z. F.; Shan, W. J.; Shen, W. J.; Wang, J. G. J. Catal. 2004 ,225, 267.
[2] Zheng, X. C.; Zhang, X. L.; Fang, Z. Y.; Wang, X. Y.; Wang, S. R.; Wu, S. H. Catal. Commun. 2006, 7, 701.
[3] Budroni, G.; Corma, A.; Angew. Chem. Int. Ed. 2006, 45, 3328.
[4] Mariño, F.; Descorme, C.; Duprez, D. Appl. Catal. B 2004, 54, 59.
[5] Fu, Q.; Weber, A.; Flytzani-Stephanopoulos, M. Catal. Lett. 2001,77, 87.
[6] Wang, J. B.; Tsai, D. H.; Huang, T. J. J. Catal. 2002, 208, 370.
[7] Martínez-Afias, A.; Hungria, A. B.; Fernández-García, M.; Conesa, J. C.; Munuera, G. J. Phys. Chem. B 2004, 108, 17983.
[8] Sedmak, G.; Hočevar, S.; Levec, J. J. Catal. 2003, 213, 135.
[9] Liu, W.; Flytzani-Stephanopoulos, M. J. Catal. 1995, 153, 304.
[10] Wang, X. Q.; Rodriguez, J. A.; Hanson, J. C.; Gamarra, D.; Martínez-Arias, A.; Fernández-García, M. J. Phys. Chem. B 2006, 110, 428.
[11] Shan, W. J.; Shen, W. J.; C. Li, Chem. Mater. 2003, 15, 4761.
[12] Avgouropoulos, G.; Ioannides, T.; Papadopoulou, C.; Batista, J.; Hocevar, S.; Matralis, H. K. Catal. Today 2002, 75, 157.
[13] Avgouropoulos, G.; Loannides, T. Appl. Catal., A 2003, 244, 155.
[14] Phonthammachai, N.; Rummangwonga, M.; Gularib, E.; Jamiesonc, A. M.; Jitkamka, S.; Wongkasemjit, S. Colloids Surfaces A. 2004, 247, 61.
[15] Shen, W. H.; Dong, X. P.; Zhu, Y. F.; Chen, H. R.; Shi, J. L. Microporous Mesoporous. Mater. 2005, 85, 157.
[16] Luo, M. F.; Ma, J. M.; Lu, J. Q.; Song, Y. P.; Wang, Y. J. J. Catal. 2007, 246, 52.
[17] Xie, G. Q.; Luo, M. F.; He, M.; Fang, P.; Ma, J. M.; Ying, Y. F.; Yan, Z. L.; J. Nanopart. Res. 2007, 9, 471.
[18] Luo, M. F.; Zhong, Y. J.; Yuan, X. X.; Zheng, X. M. Appl. Catal. A 1997, 162, 121.
[19] Luo, M. F.; Song, Y. P.; Wang, X. Y.; Xie, G. Q.; Pu, Z. Y.; Fang, P.; Xie, Y. L. Catal. Commun. 2007, 8, 834.
[20] Luo, M. F.; Pu, Z. Y.; He, M.; Jin, J.; Jin, L. Y.; J. Mol. Catal. A 2006, 260, 152.
[21] Young, R. A. The Rietveld Method, Oxford University Press, Oxford, UK, 1993; p. 22.
[22] Lutterotti, L.; Gialanella, S. Acta Mater. 1998, 46, 101.
[23] Bumajdad, A.; Zaki, M. 1.; Eastoe, J.; Pasupulety, L. Langmuir. 2004, 20, 11223.

[24] Yao, H. C.; Yu, Yao, Y. F. J. Catal. 1984, 86, 254.
[25] Tang, X. L.; Zhang, B. C.; Li, Y.; Xu, Y. D.; Xin, Q.; Shen, W. J. Catal. Today 2004, 93-95, 191.
[26] Avgouropoulos, G.; Ioannides, T. Appl. Catal. B 2006, 67, 1.
[27] Shan, W. J.; Feng, Z. C.; Li, Z. L.; Zhang, J.; Shen, W. J.; Li, C. Catal. 2004, 228, 206.
[28] Zou, H. B.; Dong, X. F.; Lin, W. M. Appl. Surf. Sci. 2006, 253, 2893.
[29] Shiau, C. Y.; Ma, M. W.; Chuang, C. S. Appl. Catal. A 2006, 301, 89.
[30] Wang, J. B.; Shih, W. H.; Huang, T. J. Appl. Catal. A 2000, 203, 191.
[31] Martínez-Arias, A.; Fernández-García, M.; Soria, J.; Conesa, J. C. J. Catal. 1999, 182, 367.
[32] Martínez-Arias, A.; Femández-García, M.; Gálvez, O.; Coronado, J. M.; Anderson, J. A.; M.; Conesa, J. C.; Soria, J.; Munuera, G. L Catal. 2000, 195, 207.
[33] Skårman, B.; Grandjean, D.; Benfield, R. E.; Hinz, A.; Andersson, A.; Wallenberg, L. R. J. Catal. 2002, 211, 119.
[34] Luo, M. F.; He, M.; Xie, Y. L.; Fang, P.; Jin, L. Y. Appl. Catal. B 2007, 69, 213.

An evidence for the strong association of N-methyl-2-pyrrolidinone with some organic species in three Chinese bituminous coals

Liu Chan-Min[1], Zong Zhi-Min[1], Jia Ji-Xian[1], Liu Guang-Feng[1] & Wei Xian-Yong[1,2,3]

[1] School of Chemical Engineering, China University of Mining and Technology, Xuzhou 221008, China;
[2] School of Life Sciences, Xuzhou Normal University, Xuzhou 221116, China;
[3] Key Laboratory of Coal Conversion and New Carbon Materials, Hubei Province, Wuhan University of Science and Technology, Wuhan 430081, China

Three Chinese bituminous coals collected from Shenfu, Heidaigou and Feicheng coal fields were respectively extracted with carbon-disulfide/N-methyl-2- pyrrolidinone (CS_2/NMP) mixed solvent (volume ratio 1 : 1) at room temperature followed by distillation of CS_2 under ambient pressure and subsequent removal of most of NMP by distillation at 110 ℃ under reduced pressure to afford mixed solvent-extractable fractions (MSEFs) with small amount of NMP. Acetone-extractable fraction 1 (AEF1) was obtained by extracting each MSEF under ultrasonic irradiation at room temperature and subsequently using a Soxhlet extractor. Direct extraction of each bituminous coal affords acetone-soluble fraction 2 (AEF2). GC/MS analysis shows that m/z of base or secondary peak in mass spectra of a series of components from each AEF1 is 98, whereas such components were not detected in AEF2. Since m/z of base peak in mass spectrum of NMP itself is 99, the base or secondary peak at m/z 98 should result from loss of —H from NMP, i.e., NMP strongly associates some organic species (OSs) and thereby the components detected with base or secondary peak at m/z 98 in their mass spectra should be associated NMP-OS.

---

① Supported by the National Natural Science Foundation of China (Grant Nos. 90410018, 90510008 and 10432060), the Key Project of Chinese Ministry of Education (Grant No. 104031) and the Program of the Universities in Jiangsu Province for Development of High—Tech Industries (Grant No. JHB05—33).

An evidence for the strong association of N-methyl-2-pyrrolidinone with some organic species in three Chinese bituminous coals

LIU ChanMin[1,2], ZONG ZhiMin[1], JIA JiXian[1], LIU GuangFeng[1] & WEI XianYong[1,3†]

[1] School of Chemical Engineering, China University of Mining and Technology, Xuzhou 221008, China;
[2] School of Life Sciences, Xuzhou Normal University, Xuzhou 221116, China;
[3] Key Laboratory of Coal Conversion and New Carbon Materials of Hubei Province, Wuhan University of Science and Technology, Wuhan 430081, China

Three Chinese bituminous coals collected from Shenfu, Heidaigou and Feicheng coal fields were respectively extracted with carbon-disulfide/N-methyl-2-pyrrolidinone (CS_2/NMP) mixed solvent (volume ratio 1:1) at room temperature followed by distillation of CS_2 under ambient pressure and subsequent removal of most of NMP by distillation at 110℃ under reduced pressure to afford mixed solvent-extractable fractions (MSEFs) with small amount of NMP. Acetone-extractable fraction 1 (AEF1) was obtained by extracting each MSEF under ultrasonic irradiation at room temperature and subsequently using a Soxhlet extractor. Direct extraction of each bituminous coal affords acetone-soluble fraction 2 (AEF2). GC/MS analysis shows that m/z of base or secondary peak in mass spectra of a series of components from each AEF1 is 98, whereas such components were not detected in AEF2. Since m/z of base peak in mass spectrum of NMP itself is 99, the base or secondary peak at m/z 98 should result from loss of α-H from NMP, i.e., NMP is strongly associated with some organic species (OSs) and thereby the components detected with base or secodary peak at m/z 98 in their mass spectra should be associated NMP-OS.

N-methyl-2-pyrrolidinone, bituminous coals, association, GC/MS analysis

Iino et al.[1,2] found that carbon disulfide/N-methyl-2-pyrrolidinone (CS_2/NMP) mixed solvent (volume ratio 1:1) is excellent for dissolving some bituminous coals. This finding has attracted great attention of coal researchers[3−6]. Investigation on thermal reaction of CS_2 with NMP reported by Zong et al. shows that there is a strong π–π interaction between C=S bond in CS_2 and C=O bond in NMP and association of the associated species with organic matter in coals was presumed to be an important reason for enhancing coal solubility in the mixed solvent[7,8]. The results from calculation based on quantum chemistry also verified the π–π interaction between CS_2 and NMP[9,10]. However, to our knowledge, there have been no reports on direct evidence for the strong interaction between organic matter in coals and associated CS_2–NMP or NMP alone.

We extracted three Chinese bituminous coals directly and CS_2/NMP mixed solvent-extractable fractions (MSEFs) of the coals with acetone, respectively, and compared the differences in composition between the resulting acetone-extractable fractions (AEFs) by GC/MS analysis, finding that AEF1 (i.e. AEF resulting from each MSEF) contains a series of components with base or secondary peak at m/z 98 in their mass spectra, whereas there are no components in AEF2 (i.e. AEF di-

---

Received May 11, 2007; accepted October 16, 2007
doi: 10.1007/s11434-007-××××-1
†Corresponding author (email: wei_×××××××@163.com)
Supported by the National Natural Science Foundation of China (Grant Nos. 90410018, 90510008 and 10432060), the Key Project of Chinese Ministry of Education (Grant No. 104031) and the Program of the Universities in Jiangsu Province for Development of High-Tech Industries (Grant No. JHB05-33)

N-methyl-2-pyrrolidinone, bituminous coals, association, GC/MS analysis

Iino et al. found that carbon disulfide/N-methyl-2-pyrrolidinone (CS_2/NMP) mixed solvent (volume ratio 1 : 1) is excellent for dissolving some bituminous coals[1,2]. This finding has attracted great attention of coal researchers[3-6]. Investigation on thermal reaction of CS_2 with NMP reported by Zong et al. shows that there is a strong interaction between C=S bond in CS_2 and C=O bond in NMP and association of the associated species with organic matter in coals was presumed to be an important reason for enhancing coal solubility in the mixed solvent[7,8]. The results from calculation based on quantum chemistry also proved the interaction between CS_2 and NMP[9,10]. However, to our knowledge, there are no reports on direct evidence for the strong interaction between organic matter in coals and associated CS_2—NMP or NMP alone.

We extracted three Chinese bituminous coals directly and CS_2/NMP mixed solvent-extractable fractions (MSEFs) of the coals with acetone, respectively, and compared the difference in composition among the resulting acetone-extractable fractions (AEFs) by GC/MS analysis, finding that AEF1 (i.e., AEF resulting from each MSEF) contains a series of components with base or secondary peak at m/z 98 in their mass spectra, whereas there are no the components in AEF2 (i.e., AEF directly resulting from each coal sample). In this paper, origin of the fragmental ion (FI) in mass spectra obtained is discussed and the FI is pointed out to be an evidence for the strong association of NMP with some organic species in these bituminous coals.

1. Experiments

1.1. Instruments and facilities

Main facilities for coal extraction and concentration of the extraction solution are CQ50 ultrasonic generator made by Shanghai Medical Instrument Plant, Hitachi himac CR 22E centrifuge, Soxhlet extractor manufactured by Mingxing Glass Instrument Plant of Jianhu County, Jiangsu Province and Büchi R-13 rotary evaporator. The instrument for analysis of extraction solution is Hewlett Packard 6890/5973 GC/MS.

1.2. Coal samples and reagents

The coals used were collected from Shenfu (Shaanxi), Heidaigou (Inner Mongolia) and Feicheng (Shandong) coal fields. They are denoted as SFC,

HDGC and FCC, respectively, for the convenience of description. After being pulverized and grinded to pass through a 200-mesh sieve (particle size < 75 m), the coals were dried in vacuum at 80 ℃ for 24 h, cooled to room temperature in vacuum. Before use, the coal samples were preserved in a desiccator. Table 1 lists data of their proximate and ultimate analyses. Solvents CS_2, NMP and acetone used are analytical pure reagents.

Table 1 Proximate and ultimate analyses (W%) of the coal samples

Coal sample	Proximate analysis			Ultimate analysis (daf)			$S_{t,d}$	H/C
	M_{ad}	A_d	V_{daf}	C	H	N		
SFC	10.2	6.5	37.7	80.5	4.8	0.9	0.4	0.7106
HDGC	6.1	23.2	37.8	79.6	4.7	1.4	0.6	0.7036
FCC	3.0	21.8	32.6	79.9	4.6	1.1	0.6	0.6861

1.3. Extraction of the coal samples and GC/MS analysis of the extraction solutions

A coal sample (5 g), NMP (250 mL) and CS_2 were sequentially added into a 1000 mL conical flask. After 2 h extraction under ultrasonic irradiation (UI), the mixture including extraction solution and residue in the flask was transferred into some tubes used for centrifugation and centrifugated at 10000 rpm for 10 min. The extraction solution in the tubes was filtrated through a ploytetraflu oroethylene membrane filter (PMF) with a pore size of 0.8 m and the residue was extracted with the same amount of CS_2/NMP mixed solvent as mentioned above. The above operation was repeated 12 times. All the filtrates were incorporated and distillated under ambient pressure to remove CS_2 followed by distillation at 110 ℃ under reduced pressure with a rotary evaporator to remove most of NMP. MSEF with small amount of NMP was obtained thereby. The CS_2/NMP insoluble residue was repeatedly subject to UI in acetone and the slurry was filtrated through the same type of PMF until no NMP can be detected from filtrate. The NMP free residue was dried in vacuum at 80 ℃ for 24 h and cooled to room temperature in vacuum before weighing. The MSEF yield Y was calculated by dry-base weight $W_{R,d}$ of the dried NMP-free residue according the following formula:

$$Y = (W_d - W_{R,d})/W_{daf}$$

where W_d and W_{daf} denote dry and organic-base weights of the coal sample, respectively.

The MSEF was extracted with 300 mL of acetone under UI for 2 h and the resulting slurry was separated to filter cake and filtrate by filtration through the same type of PMF. Then, the filter cake was transferred into a Soxhlet extractor and extracted with acetone for 10 days. The filtrate and extraction solution were incorporated as a solution consisting of acetone, acetone-extractable fraction 1 (AEF1) and small amount of NMP. Another solution, which contains acetone and acetone-extractable fraction 2 (AEF2) was also obtained by directly extracting each coal with acetone according to similar procedure. The above solutions were analyzed with GC/MS. The residues from each MSEF and coal sample were dried in vacuum at 80 ℃ for 24 h and cooled to room temperature before weighing. The yields of AEF1 and AEF2 were calculated based on the residue weights.

2. Results and Discussion

2.1. Comparison of the yields of MSEF, AEF1 and AEF2

As Figure 1 shows, the yields of MSEF, AEF1 and AEF2 decrease in the order: FCC (33.5%) > SFC (30.8%) > HDGC (20.4%), SFC

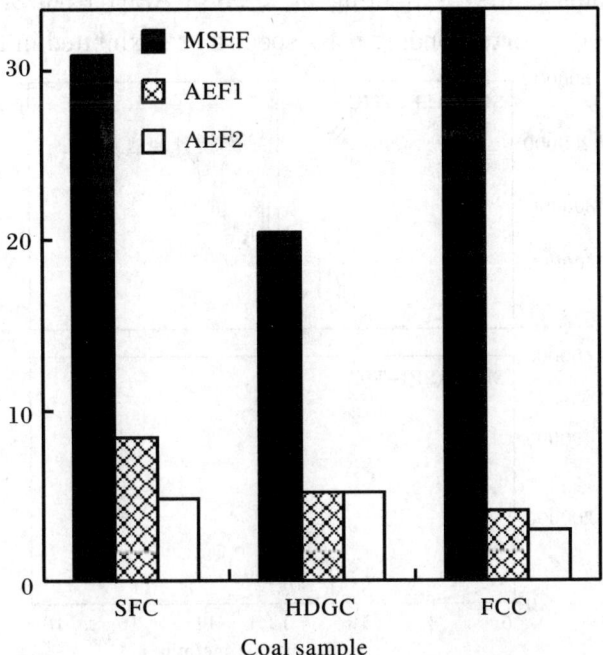

Figure 1 Yields of MSEF, AEF1 and AEF2 from three bituminous coals.

(8.3%) > HDGC (5.2%) > FCC (4.1%) and HDGC (5.1%) > SFC (4.8%) > FCC (2.9%), respectively. As listed in Table 1, the three coal samples have almost the same contents of C and H and the differences among their H/C atomic ratio are not significant either, but the differences among the yields of the extracts from different coal samples are remarkable. Although the yield of MSEF is the highest, those of AEF1 and AEF2 are the lowest from FCC, which should be ascribed to the differences in composition and structure among the coals. The yield of AEF1 is higher than that of AEF2 from the same coal sample. Probable reason for the result is that for direct extraction of a coal sample, a part of acetone soluble components (ASCs) were wrapped into acetone-insoluble components, leading to "capsule effect". CS_2/NMP mixed solvent can dissolve the "capsule" so that ASCs wrapped in the "capsule" can be released.

2.2. GC/MS analysis of AEF1 and AEF2

We unexpectedly detected a series of components with base or secondary peak at m/z 98 in their mass spectra during GC/MS analysis of AEF1 from each coal sample. Figures 2 to 4 show total ion chromatograms (TICs) and selective ion chromatograms (SICs) extracting m/z 98 of AEF1 from SFC, HDGC and FCC, respectively. Corresponding mass spectra are exhibited in Figures 5 to 7.

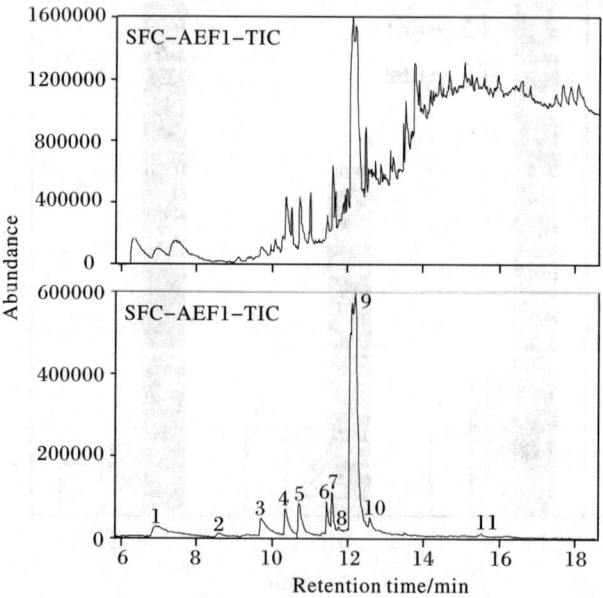

Figure 2 TIC (upper) and SIC extracting m/z 98 (lower) of AEF1 from SFC.

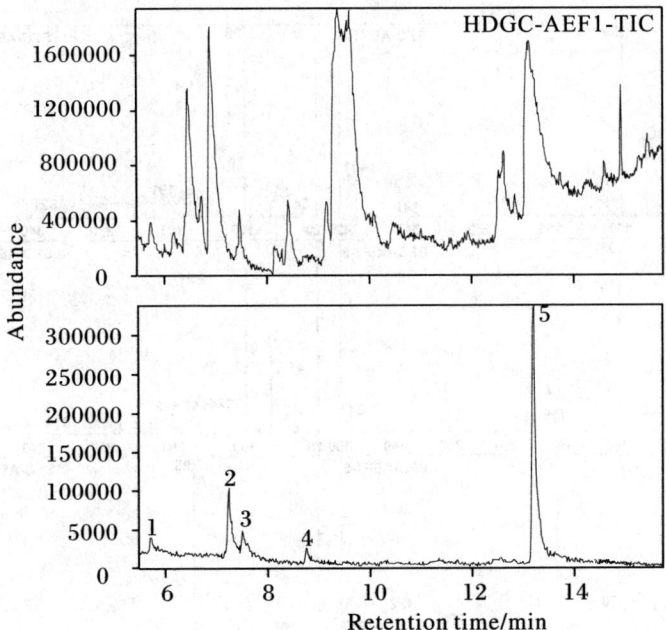

Figure 3 TIC (upper) and SIC extracting m/z 98 (lower) of AEF1 from HDGC.

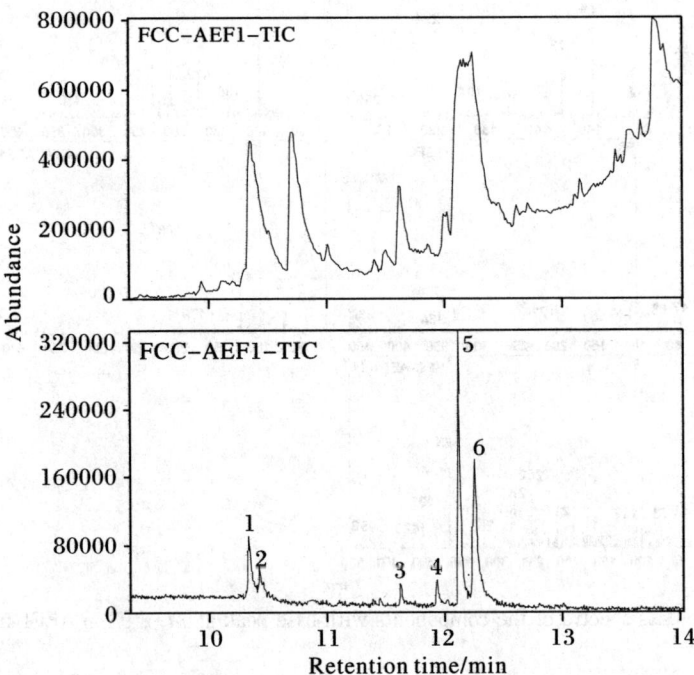

Figure 4 TIC (upper) and SIC extracting m/z 98 (lower) of AEF1 from FCC.

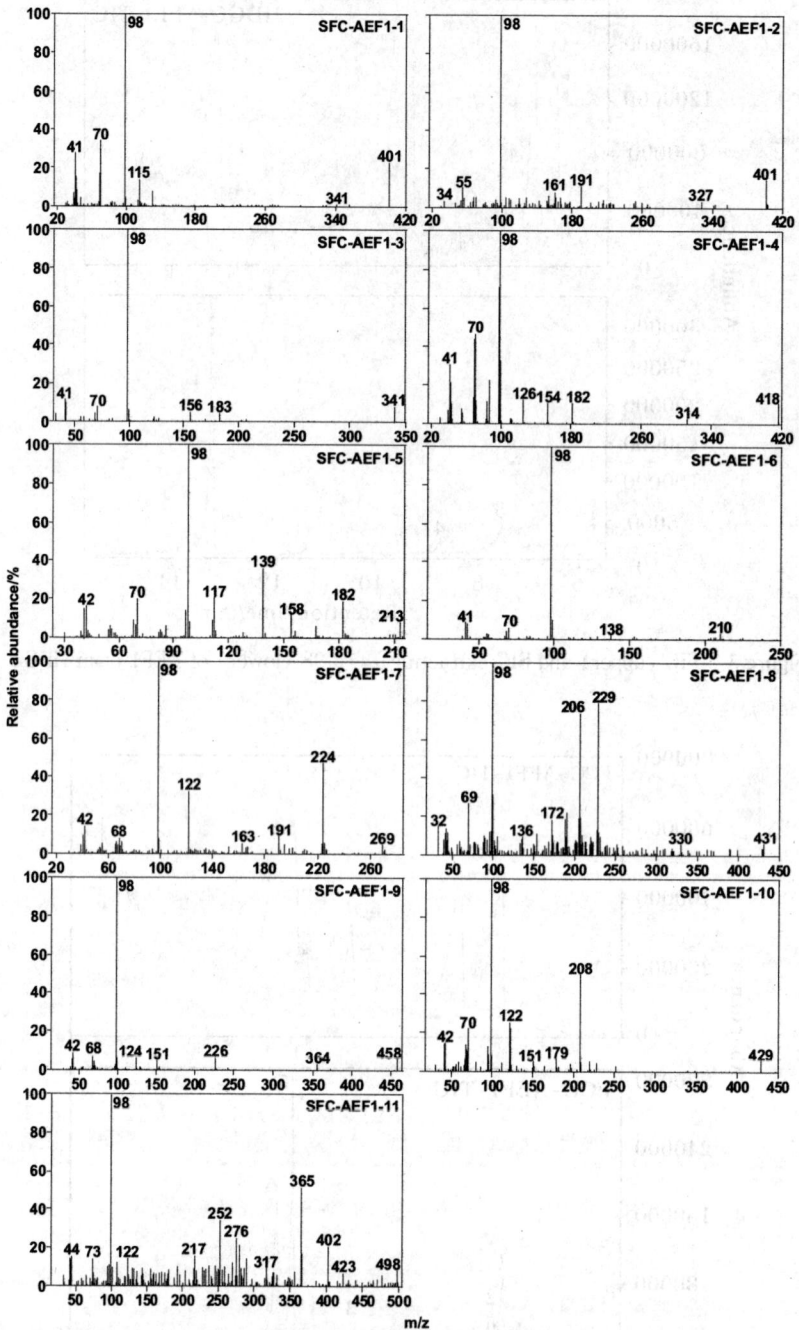

Figure 5 Mass spectra of the components with base peak at m/z 98 in AEF1 from SFC.

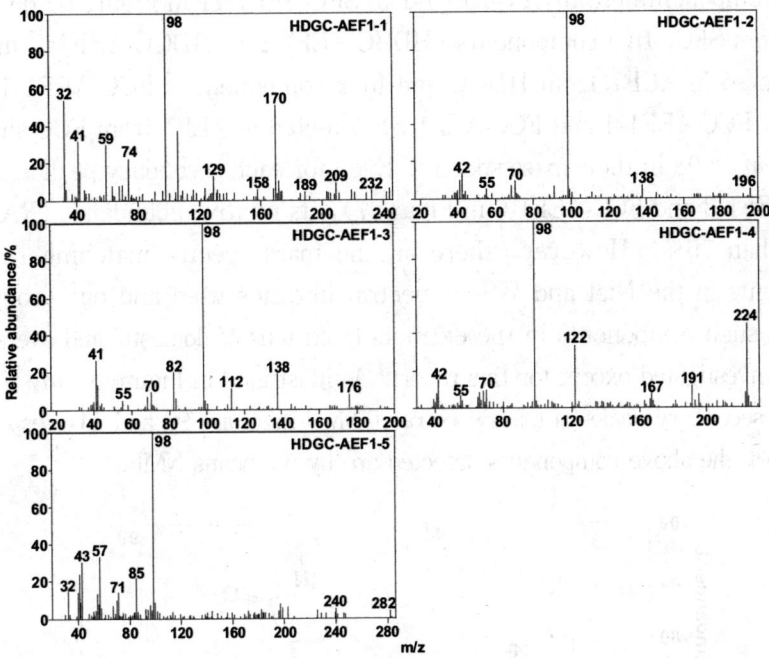

Figure 6 Mass spectra of the components with base peak at m/z 98 in AEF1 from HDGC.

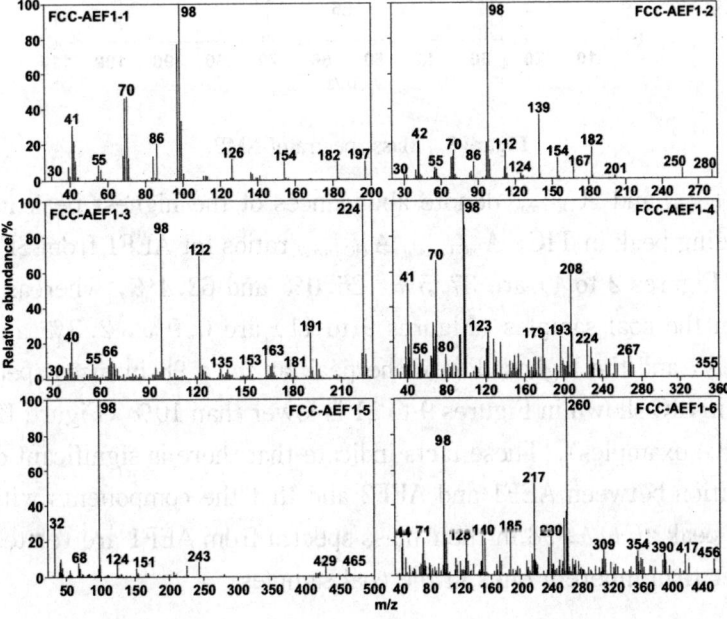

Figure 7 Mass spectra of the components with base or secondary peak at m/z 98 in AEF1 from FCC.

Eleven components (from SFC-AEF1-1 to SFC-AEF1-11 in Figure 5) detected in AEF1 from SFC, five components (HDGC-AEF1-1 to HDGC-AEF1-5 in Figure 6) detected in AEF1 from HDGC and four components (FCC-AEF1-1, FCC-AEF1-2, FCC-AEF1-4 and FCC-AEF1-5) detected in AEF1 from FCC show base peak at m/z 98 in their mass spectra. Even for each secondary peak at m/z 98 (FCC-AEF1-3 and FCC-AEF1-6 in Figure 7), its relative abundance (RA) is still higher than 75%. However, there are no mass spectra matching the above components in the Nist and Wiley spectral libraries used and our group never detected such components in the extracts from tens of domestic and overseas coal samples investigated except for this report. As illustrated in Figure 8, m/z ratios of base and secondary peaks in mass spectra of NMP itself are 99 and 44, respectively. Therefore, the above components detected are by no means NMP.

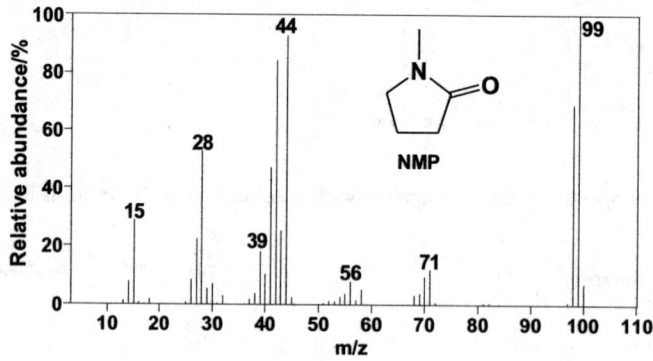

Figure 8 Mass spectra of NMP.

If $A_{\text{SIC (max)}}$ and $A_{\text{TIC (max)}}$ denote abundances of the highest peak in SIC and corresponding peak in TIC, $A_{\text{SIC (max)}}/A_{\text{TIC (max)}}$ ratios for AEF1 from SFC, HDGC and FCC (Figures 2 to 4) are 37.5%, 26.0% and 62.4%, whereas those for AEF2 from the coal samples (Figures 9 to 11) are 0.9%, 2.3% and 1.8%, respectively, and the highest RA of the peak at m/z 98 in mass spectra of all the peaks in SIC shown in Figures 9 to 11 is lower than 10% (Figure 12 displays representative examples). These facts indicate that there is significant difference in composition between AEF1 and AEF2 and that the components with base or secondary peak at m/z 98 in their mass spectra from AEF1 are related to NMP used rather than inherent ones in the coal samples.

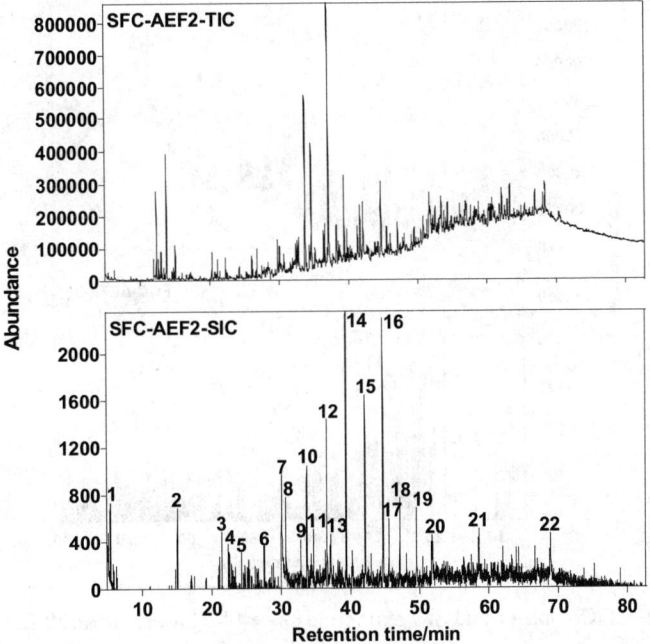

Figure 9 TIC (upper) and SIC extracting m/z 98 (lower) of AEF2 from SFC

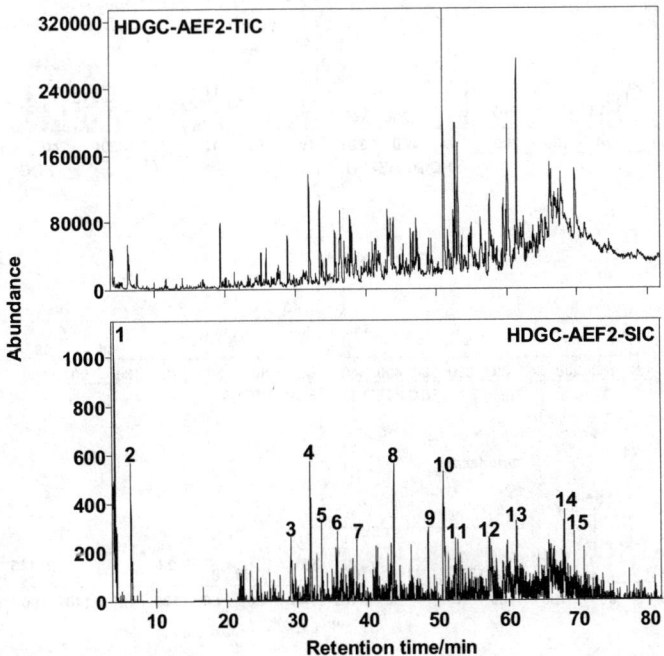

Figure 10 TIC (upper) and SIC extracting m/z 98 (lower) of AEF2 from HDGC.

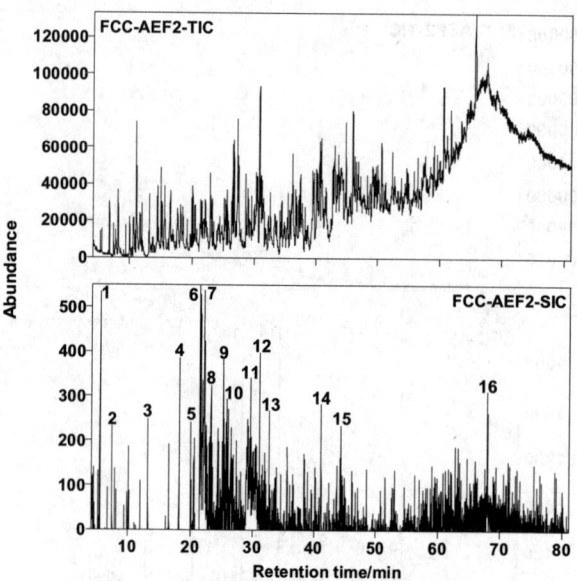

Figure 11 TIC (upper) and SIC extracting m/z 98 (lower) of AEF2 from FCC

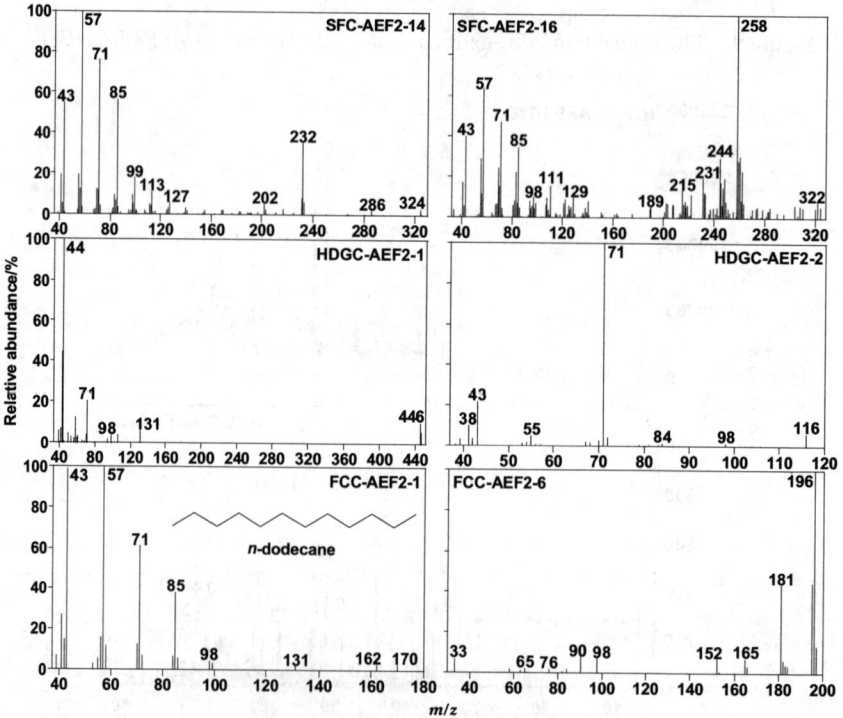

Figure 12 Mass spectra of two components containing m/z 98 with the largest abundances in AEF2 from each coal sample.

2.3. Examination of the mechanism for the formation of components with base or secondary peak at m/z 98 in their mass spectra

Taking data in Figures 5 to 7 into account, m/z ratios of molecular ions of the components with base or secondary peak at m/z 98 in their mass spectra are much larger than 98, i.e., the base and secondary peaks are FI peaks of the components. Since molecular mass of NMP is 98 and NMP contains —H, FI with base or secondary peak at m/z 98 should result from the loss of —H from NMP. However, as shown in Figure 8, m/z ratio of neither base nor secondary peak in mass spectrum of NMP is 98, indicating that the loss of —H from NMP alone is not easy during GC/MS analysis and that the components with base or secondary peak at m/z 98 in their mass spectra detected in AEF1 are nothing but some products formed by association of NMP with some OSs in the coals, i.e., associated NMP—OS.

As Figure 13 illustrates, if association between NMP and OS is not strong, the associated NMP—OS is likely to thermally dissociate at once after being injected into gasification chamber of the GC during GC/MS analysis; if the association is strong enough, the associated NMP—OS may be able to enter mass detector through capillary column in the GC and is ionized, forming M_1^+ or M_2^+. FI with base or secondary peak at m/z 98 can be detected only in the case of the M_1^+ formation. In other words, the experimental fact that a series of components with base or secondary peak at m/z 98 in their mass spectra were detected in AEF1 from three Chinese bituminous coals provides an evidence for strong association of NMP with some OSs in the bituminous coals.

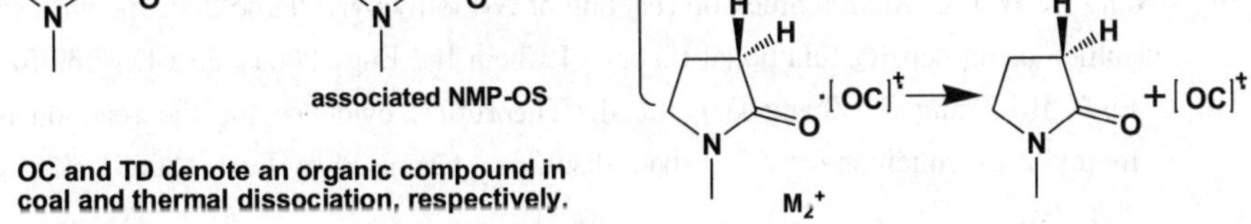

Figure 13 Mechanism for dissociation of associated NMP—OS during GC/MS analysis.

3. Summary

A series of components with base or secondary peak at m/z 98 in their mass spectra were detected in AEF1 by GC/MS analysis but such components were not detected in AEF2 from three Chinese bituminous coals. In addition, m/z ratios of base and secondary peaks in mass spectra of NMP are not 98. These facts indicate that there are strong associations between NMP and some OSs in the coals.

References

[1] Iino M, Kumagai J, Ito O. Coal extraction with carbon-disulfide mixed solvent at room temperature. J. Fuel Soc Jpn, 1985, 64 (3): 210-212.

[2] Iino M, Takanohashi T, Ohsuga H, et al. Extraction of coals with CS_2-N-methyl-2-pyrrolidinone mixed solvent at room temperature: Effect of coal bank and synergism of the mixed solvent. Fuel, 1988, 67 (12): 1639-1647.

[3] Cai M F, Smart R B. Quantitative analysis of N-methyl-2-pyrrodinone in coal extracts by TGA-FTIR. Energy Fuels, 1993, 7 (1): 52-56.

[4] Mochida I, Kinya S. Advances in Catalysis. New York: Academic Press, 1994.

[5] Chervenick S W, Smart R B. Quantitative analysis of N-methyl-2-pyrrolidinone retained in coal extracts by thermal extraction G. C.-M. S. Fuel, 1995,74 (2): 241-245.

[6] Gao H, Nomura M, Murata S, et al. Statistical distribution characteristics of pyridine transport in coal particles and a series of new phenomenological models for overshoot and nonovershoot solvent swelling of coal particles. Energy Fuels, 1999, 13 (2): 518-528.

[7] Zong Z M, Peng Y L, Qin Z H, et al. Reaction of N-methyl-2-pyrrodinone with carbon disulfide. Energy Fuels, 2000,14 (3): 734-735.

[8] Zong Z M, Peng Y L, Liu Z G, et al. Convenient synthesis of N-methylpyrrolidine-2-thione and some thioamides. Korean J Chem Eng, 2003, 20 (2): 235-238.

[9] Wang B, Wei X, Xie K. Study on reaction of N-methylpyrrolidine-2-thione with carbon disulfide using density functional theory. J Chem Ind Eng, 2004, 55 (4): 569-574.

[10] Fu X B, Zhang C, Zhang D J, et al. Theoretical evidence for the reaction of N-methyl-2-pyrrolidinone with carbon disulfide. Chem Phys Lett, 2006, 420 (1): 162-165.

Acrylonitrile synthesis from acetonitrile and methanol over MgMn/ZrO$_2$ catalysts

Leihong Zhao*, Jianjun Zhao, Mengfei Luo, Guangyu Liu, Jie Song, Yongping Zhang

Institute of Physical Chemistry, Zhejiang Key Laboratory for Reactive Chemistry on Solid Surfaces, Zhejiang Normal University, Jinhua 321004, China

Received 26 January 2005; accepted 11 May 2005

Available online 5 July 2005

Abstract

A series of MgMn/ZrO$_2$ catalysts with different Mn loadings were prepared by an impregnation method. The catalysts were used for the synthesis of acrylonitrile from acetonitrile and methanol. The addition of a suitable amount of Mn exhibited promoting effects on both the activity and the selectivity to acrylonitrile formation. The optimum Mn-content was determined to be Mn/Zr = 0.2 (molar ratio), over which the acrylonitrile yield could reach 13.6%. Based on various characterizations, such as XRD, XPS, SEM, CO$_2$—TPD and nitrogen physisorption, the promoting effect of the Mn-dopant on the catalytic property was mainly attributed to the presence of Mg$_2$MnO$_4$ and the strong(ly) basic site. However, at high Mn content, the activity decreased due to the coverage of the active sites by inactive Mn$_2$O$_3$ and the weak basicity of the catalysts.

Keywords: MgMn/ZrO$_2$ catalyst, acrylonitrile, acetonitrile, methanol

1. Introduction

Acrylonitrile is one of the most important chemicals in chemical industries, especially in polymer industries [1,2]. Generally, it is synthesized by the ammoxidation of propylene over multicomponent bismuth molybdate catalysts[3–6].

Acrylonitrile synthesis from acetonitrile and methanol over MgMn/ZrO$_2$ catalysts

Leihong Zhao [*], Jianjun Zhao, Mengfei Luo, Guangyu Liu, Jie Song, Yongping Zhang

Institute of Physical Chemistry, Zhejiang Key Laboratory for Reactive Chemistry on Solid Surfaces, Zhejiang Normal University, Jinhua 321004, China

Received 26 January 2005; accepted 11 May 2005

Available online 5 July 2005

Abstract

A series of MgMn/ZrO$_2$ catalysts with different Mn loadings were prepared by an impregnation method. The catalysts were used for the synthesis of acrylonitrile from acetonitrile and methanol. The addition of a suitable amount of Mn exhibited promoting effects on both the activity and the selectivity to acrylonitrile formation. The optimum Mn-content was determined to be Mn/Zr = 0.2 (molar ratio), over which the acrylonitrile yield could reach 13.6%. Based on various characterizations, such as XRD, XPS, SEM, CO$_2$-TPD and nitrogen physisorption, the promoting effect of the Mn-dopant on the catalytic property was mainly attributed to the presence of Mg$_2$MnO$_4$ and the strong(ly) basic site. However, at high Mn content, the activity decreased due to the coverage of the active sites by inactive Mn$_2$O$_3$ and the weak basicity of the catalysts.

© 2005 Elsevier B.V. All rights reserved.

Keywords: MgMn/ZrO$_2$ catalyst; Acrylonitrile; Acetonitrile; Methanol

1. Introduction

Acrylonitrile is one of the most important chemicals in chemical industries, especially in polymer industries [1,2]. Generally, it is synthesized by the ammoxidation of propylene over multicomponent bismuth molybdate catalysts [3–6]. The main by-product during the production of acrylonitrile is acetonitrile. Although acetonitrile can undergo typical nitrile reactions leading to amines, amides, halogenated nitriles, ketones, and others [3,7], the applicable amount is little. Acetonitrile is used mainly as a solvent or it is incinerated as its supply far exceeds by its demand. Hence, the use of acetonitrile for the production of more valuable chemicals and products will be a very important technological achievement.

In recent years, several investigators have tried to add a carbon atom to the α-position of acetonitrile and thus convert it to acrylonitrile [3,7–13]. Formaldehyde [8], methanol [7–11], and methane [3,12,13] have been utilized as methylating agents. Methane is very stable and therefore higher temperature is required for its activation [3,12,13]. Methanol is a very good candidate for methylation [7–11] due to its high activity and abundant supply transformed from natural gas. Since the acetonitrile molecule possesses an electron withdrawing group (–CN), basic catalysts are required to activate the α-carbon of acetonitrile and therefore initiate the desired methylation reaction. Alkali metals supported on silica gel [8], alkali or alkaline earth metals supported on silica [13], and transition metal-promoted MgO [7,9,11] were used for the acrylonitrile synthesis from methanol and acetonitrile. The performance of the above-mentioned catalysts increased in the order Li > Na > K > Rb but the yield of acrylonitrile remained very low [8]. MgO loaded with metals such as Cr, Fe, Mn, and Cu was found to selectively transform

---

[*] Corresponding author. Tel.: +86 5792282402; fax: +86 5792282521.
E-mail address: ×××××××@163.com (L. Zhao).

1566-7367/$ - see front matter © 2005 Elsevier B.V. All rights reserved.
doi:10.1016/j.catcom.2005.05.002

The main by-product during the production of acrylonitrile is acetonitrile. Although acetonitrile can undergo typical nitrile reactions leading to amines, amides, halogenated nitriles, ketones, and others [3,7], the applicable amount is little. Acetonitrile is used mainly as a solvent or it is incinerated as its supply far exceeds by its demand. Hence, the use of acetonitrile for the production of more valuable chemicals and products will be a very important technological achievement.

In recent years, several investigators have tried to add a carbon atom to the α-position of acetonitrile and thus convert it to acrylonitrile[3, 7—13]. Formaldehyde[8], methanol [7—11], and methane [3, 12, 13] have been utilized as methylating agents. Methane is very stable and therefore higher temperature is required for its activation [3, 12,13]. Methanol is a very good candidate for methylation [7—11] due to its high activity and abundant supply transformed from natural gas. Since the acetonitrile molecule possesses an electron withdrawing group (—CN), basic catalysts are required to activate the α-carbon of acetonitrile and therefore initiate the desired methylation reaction. Alkali metals supported on silica gel [8], alkali or alkaline earth metals supported on silica [13], and transition metal-promoted MgO [7,9,11] were used for the acrylonitrile synthesis from methanol and acetonitrile. The performance of the above mentioned catalysts increased in the order Li > Na > K > Rb but the yield of acrylonitrile remained very low [8]. MgO loaded with metals such as Cr, Fe, Mn, and Cu was found to selectively transform methanol and acetonitrile to acrylonitrile at temperatures in the range 350—400 ℃ [7]. However, the maximum conversion of acetonitrile reported was about 10%. Therefore, it is necessary to develop a novel catalyst with high activity and selectivity for this reaction.

In this paper we report a series of Mn-modified magnesia supported on ZrO_2 that are used as catalysts for the synthesis of acrylonitrile from methanol and acetonitrile. These catalysts are seen to exhibit excellent activity and selectivity. According to the studies of the texture and structural property of these catalysts, the effect of Mn content on the catalytic property is also discussed.

2. Experimental

2.1. Catalyst preparation

The Mg and Mn supported on ZrO_2 catalysts were prepared by impregnation of ZrO_2 with an aqueous solution of $Mg(NO_3)_2$ and $Mn(NO_3)_2$ for 2.0 h. The suspension was kept at 95 ℃ under vigorous agitation until the water evaporated. The obtained paste was dried in the oven at 120 ℃ for 4.0 h, and then calcined in air at 600 ℃ for 5.0 h. The obtained catalysts were designated as Mg_xMn_y/ZrO_2 (x and y represent the molar ratios of Mg/Zr and Mn/Zr, respectively).

2.2. Catalytic test

The synthesis of acrylonitrile from acetonitrile and methanol was carried out in the gas phase at atmospheric pressure, using a quartz tube reactor (ϕ8 mm × 215 mm). For each catalytic experiment, approximately 500mg of catalyst was loaded into the reactor tube and placed in-between two quartz wool plugs. The catalyst was heated to reaction temperature under nitrogen flow at 70 mL/min, after which the reactants (acetonitrile and methanol with a molar ratio of 1:10 and a flow rate of 0.03 mL/min) were introduced through a heated line using a syringe pump to initiate the reaction. The reactor effluent was analyzed with an on-line 7890 II gas chromatograph equipped with a flame ionization detector (FID). Chromatograph separation was accomplished with a PEG 20M capillary column. Products obtained were acrylonitrile (AN), propionitrile (PN), and methacrylonitrile (MAN). The conversions and selectivities listed in Figure 5 to 8 were based on the nitrogen and calculated by the following equations:

$$\text{Conversion} = (([AN] + [PN] + [MAN]) / [\text{Acetonitrile}]_{\text{reactant}}) \times 100\% \quad (1)$$

$$\text{Selectivity} = ([AN] / ([AN] + [PN] + [MAN])) \times 100\%. \quad (2)$$

2.3. Characterization

The surface area was measured with the BET method by N_2 physisorption at -196 ℃ on a Micromeritics TriStar 3000 adsorption apparatus. Powder X-ray diffraction (XRD) patterns of the catalysts were determined using a Philips PW3040/60 X-ray diffractometer using Cu Kα radiation (0.15418 nm). The tube voltage and current were 40 kV and 40 mA, respectively. The surface morphology was observed by scanning electron microscopy (SEM, Philips XL 30). The X-ray photoelectron spectroscopic (XPS, Perkin-Elmer PHI 5000C) spectra were analyzed with Al Kα radiation ($h\nu$ = 1486.6 eV). All the binding energy (BE) values were referenced with regard to the C 1s peak of contaminant carbon at 284.6 eV with uncertainty of ±0.2 eV.

CO_2 desorption profiles were obtained in the following manner. After the samples were treated at 500 ℃ for 30 min under helium flow, they were cooled down to room temperature before the saturation chemisorption of carbon dioxide by pulsed injection was confirmed by the constant peak area. The maximum desorption temperature, 900 ℃, was reached at a ramping rate of 20 ℃/min. Desorbed gases were analyzed by a mass spectrometer (Omnistar 200, Balzers).

3. Results and discussion

3.1. Surface areas and morphologies of $Mg_{0.5}Mn_y/ZrO_2$ catalysts

The BET surface areas, pore volumes, and mean pore diameters of the $Mg_{0.5}Mn_y/ZrO_2$ catalysts are summarized in Table 1. The low BET surface areas of the ZrO_2 support and $Mg_{0.5}Mn_y/ZrO_2$ catalysts is due to the high calcination temperature (600 ℃)[14]. In the ZrO_2 sample, the surface area is only 16.1 m^2/g, the pore volume is 0.1463 cm^3/g, and the mean pore diameter calculated from the ratio of pore volume to area for cylindrical pores is 36.3 nm. Since Mg is supported on ZrO_2, the surface area and pore volume are about a quarter of those of the ZrO_2 sample because of the partial coverage of the pores of ZrO_2 by Mg species. The mean pore diameter from the volume-to-area ratio is about 4.6 nm, larger than that observed for ZrO_2 (because the $Mg_{0.5}/ZrO_2$ catalyst exhibited smaller surface area). However, with the increase of the Mn content, the surface areas of the $Mg_{0.5}Mn_y/ZrO_2$ catalysts increased, while the mean pore diameter decreased. A similar phenomenon has been reported for La—Cr/ZrO_2 prepared by impregnating La_2O_3/ZrO_2 with an aqueous solution of $(NH_4)_2CrO_4$ at pH = 10 [14].

Table 1 Results of nitrogen physisorption for the $Mg_{0.5}Mn_y/ZrO_2$ catalysts and ZrO_2

Catalysts	BET Surface Area (m^2/g)	Pore Volume (cm^3/g)	Pore Diameter (nm)
ZrO_2	16.1	0.1463	36.3
$Mg_{0.5}/ZrO_2$	4.6	0.0473	40.9
$Mg_{0.5}Mn_{0.05}/ZrO_2$	6.2	0.0570	36.8
$Mg_{0.5}Mn_{0.1}/ZrO_2$	6.5	0.0594	36.6
$Mg_{0.5}Mn_{0.2}/ZrO_2$	6.8	0.0582	34.2
$Mg_{0.5}Mn_{0.3}/ZrO_2$	7.1	0.0498	30.8
$Mg_{0.5}Mn_{0.4}/ZrO_2$	7.1	0.0472	26.8

Figure 1 shows the scanning electron micrographs of ZrO_2, $Mg_{0.5}/ZrO_2$,

$Mg_{0.5}Mn_{0.2}/ZrO_2$, and $Mg_{0.5}Mn_{0.4}/ZrO_2$ samples treated at 600 ℃ for 5.0 h. We can see clearly that flaky Mg species are formed on the smooth surface of the ZrO_2 support when magnesium is loaded. The thickness of these flakes was uniform

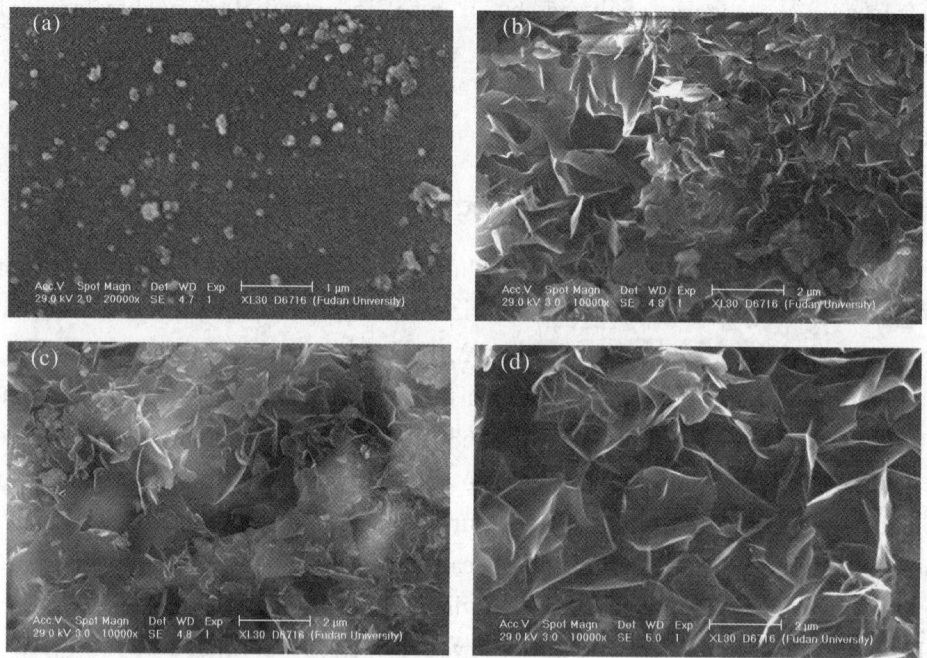

Figure 1 SEM micrographs of (a) ZrO_2, (b) $Mg_{0.5}/ZrO_2$,
(c) $Mg_{0.5}Mn_{0.2}/ZrO_2$, and (d) $Mg_{0.5}Mn_{0.4}/ZrO_2$ catalysts.

(about 40 nm), while the size ranged between 0.4 and 2.1 μm (Figure 1b). The flakes of the same thickness were also found in the $Mg_{0.5}Mn_{0.2}/ZrO_2$ catalyst (Figure 1c). However, the size of the flakes was uniform, about 0.9 μm. The larger uniform flakes appeared in the $Mg_{0.5}Mn_{0.4}/ZrO_2$ catalyst (Figure 1d).

3.2. Structure of $Mg_{0.5}Mn_y/ZrO_2$ catalysts

$Mg_{0.5}Mn_y/ZrO_2$ catalysts were calcined at 600 ℃ and analyzed with powder X-ray diffraction (shown in Figure 2). Examination of the powder diffraction patterns revealed the formation of a monoclinic ZrO_2 and cubic MgO for the $Mg_{0.5}/ZrO_2$ catalyst (Figure 2a)[15]. No diffraction lines corresponding to MgO and ZrO_2 compound was seen in Figure 2. From the XRD patterns of the Mn-modified $Mg_{0.5}/ZrO_2$ catalysts, three diffraction lines (at about 35.6°, 43.3°, and 62.8°) were found, which correspond to the three main diffraction peaks of the Mg_2MnO_4 phase[15]. A comparison of the intensities of the two lines at about

35.6° and 42.9°, corresponding to the Mg_2MnO_4 and MgO phases respectively, directly leads to the conclusion that the relative molar ratio of Mg_2MnO_4 to MgO increases with the Mn loading. It seems that the addition of manganese enhances the formation of Mg_2MnO_4 which could be considered as an active species in the synthesis of acrylonitrile from acetonitrile and methanol. When the molar ratio of Mn/Mg in the catalyst is higher than the stoichiometric ratio of Mn/Mg (1 : 2) in the Mg_2MnO_4 compound, like $Mg_{0.5}Mn_{0.3}/ZrO_2$ and $Mg_{0.5}Mn_{0.4}/ZrO_2$ catalysts, the diffraction lines of the MgO phase disappeared. Instead, the appearance of the diffraction line at about 32.9° of Mn_2O_3 phase in the XRD patterns[15] indicates that the Mn_2O_3 phase is found alongside the ZrO_2 and Mg_2MnO_4 phases in these two catalysts (Figure 2e, 2f).

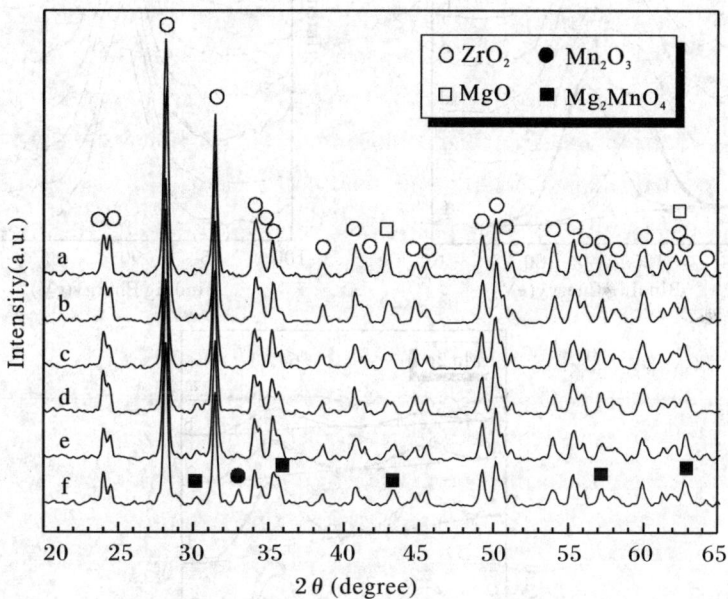

Figure 2 XRD patterns of (a) $Mg_{0.5}/ZrO_2$, (b) $Mg_{0.5}Mn_{0.05}/ZrO_2$, (c) $Mg_{0.5}Mn_{0.1}/ZrO_2$, (d) $Mg_{0.5}Mn_{0.2}/ZrO_2$, (e) $Mg_{0.5}Mn_{0.3}/ZrO_2$, and (f) $Mg_{0.5}Mn_{0.4}/ZrO_2$ catalysts.

Deraz studied a series of manganese oxides loaded on an active magnesia support which were prepared by the calcinations, at 400—800 ℃, of magnesium carbonate impregnated with manganese nitrate solution[16]. His report that the rise in calcination temperature to 600 ℃ led to the complete disappearance of MnO_2 and entire conversion into Mn_2O_3 and Mg_2MnO_4 is consistent with our work. Simultaneously, he thought that Mg_2MnO_4 was formed by solid-solid interaction between a small portion of Mn_2O_3 and magnesia according to:

$$4MgO + Mn_2O_3 + 1/2O_2 \Longrightarrow 2Mg_2MnO_4$$

3.3. XPS studies

The XPS spectra of the six catalysts are illustrated in Figure 3. For all samples, there are two peaks at the BE of 182.2 eV, and 184.6 eV, attributed

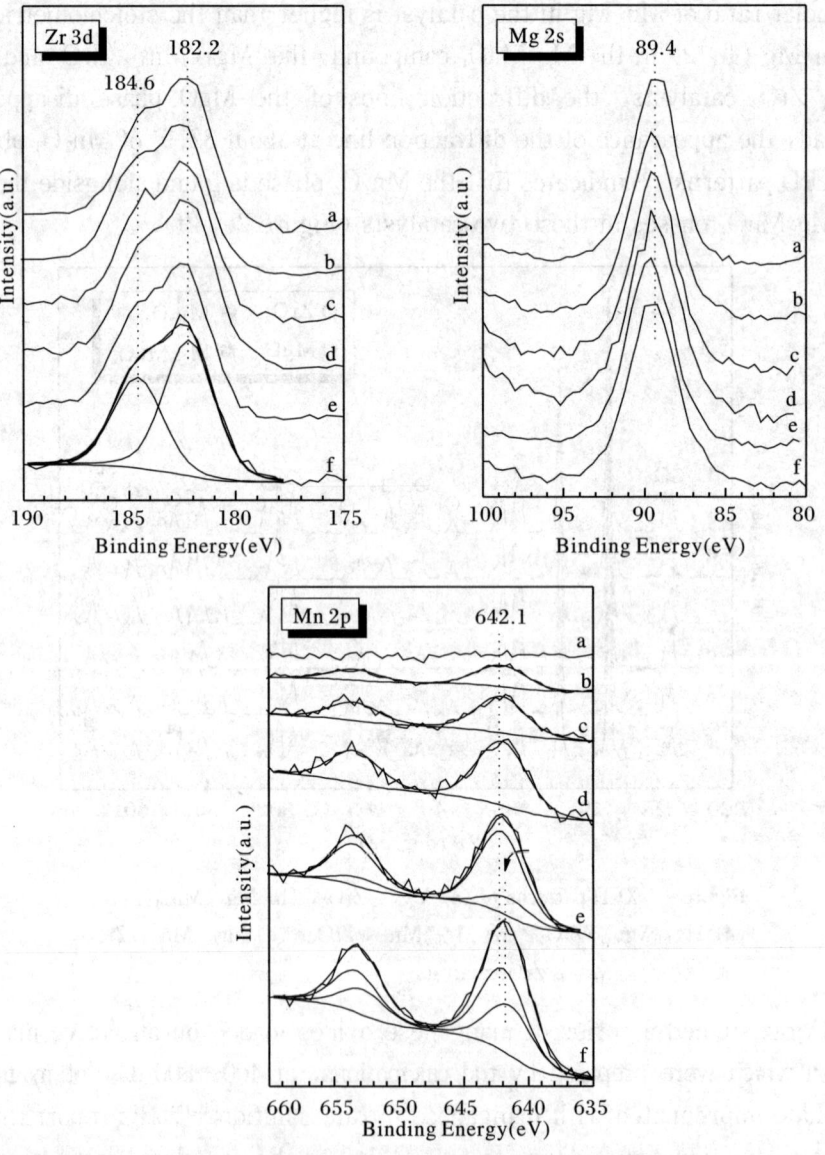

Figure 3 The Zr 3d, Mg 2s, and Mn 2p XPS spectra of (a) $Mg_{0.5}/ZrO_2$, (b) $Mg_{0.5}Mn_{0.05}/ZrO_2$, (c) $Mg_{0.5}Mn_{0.1}/ZrO_2$, (d) $Mg_{0.5}Mn_{0.2}/ZrO_2$, (e) $Mg_{0.5}Mn_{0.3}/ZrO_2$ and (f) $Mg_{0.5}Mn_{0.4}/ZrO_2$ catalysts.

to Zr $3d_{5/2}$ and Zr $3d_{3/2}$ respectively [17]. It was suggested that Zr was mainly in the ZrO_2 state in $Mg_{0.5}Mn_y/ZrO_2$ catalyst. It should be noted that the state analysis of magnesium from the Mg 2p peak was more difficult due to the interference from the Zr 4s peak and the Mn 3p peak [17]. Therefore, the Mg 2s XPS spectra were studied. The BE of Mg 2s for $Mg_{0.5}Mn_y/ZrO_2$ catalysts, 89.4 eV, confirms that Mg(Ⅱ) was the dominant species [17, 18], which is consistent with the formation of MgO and Mg_2MnO_4 in catalysts measured by XRD. In the Mn 2p region, the intensity of the Mn XPS peak clearly increases with manganese loading in the catalyst. For $Mg_{0.5}Mn_{0.05}/ZrO_2$, $Mg_{0.5}Mn_{0.1}/ZrO_2$, and $Mg_{0.5}Mn_{0.2}/ZrO_2$ catalysts, manganese is mainly in its Mg_2MnO_4 state with a Mn $3d_{5/2}$ BE of 642.1 eV [17]. However, the peaks of Mn 3d shifted to low binding energy in the $Mg_{0.5}Mn_{0.3}/ZrO_2$, and $Mg_{0.5}Mg_{0.5}Mn_{0.4}/ZrO_2$ catalysts, implying that another state of Mn is present. From curve fitting of the experimental spectra, the spectra shows two peaks positioned at 641.6 eV and 642.1 eV which are assigned to be Mn_2O_3 and Mg_2MnO_4, respectively [17]. These results further confirm that manganese in Mg_2MnO_4 is the dominant manganese species in the lower Mn/Mg ratio, while the formation of Mn_2O_3 and Mg_2MnO_4 compounds are seen in the $Mg_{0.5}Mn_y/ZrO_2$ catalysts with the higher Mn/Mg ratio.

By a simple calculation, the atomic ratio of Mg/Mn in the Mn_2O_3 and Mg_2MnO_4 phases is 0 and 2, respectively. If all Mn and Mg are in the formation of Mn_2O_3 and Mg_2MnO_4, the Mn atomic ratio of Mn_2O_3 and Mg_2MnO_4 is 1∶5 for the $Mg_{0.5}Mn_{0.3}/ZrO_2$ catalyst, and 3∶5 for the $Mg_{0.5}Mn_{0.4}/ZrO_2$ catalyst. According to the peak areas of Mn 3d shown in Figure 3, the Mn atomic ratio of Mn_2O_3 and Mg_2MnO_4 in the $Mg_{0.5}Mn_{0.3}/ZrO_2$ and $Mg_{0.5}Mn_{0.4}/ZrO_2$ catalyst was 0.22 and 0.61, respectively, close to the above calculated values.

3.4. Basicity of $Mg_{0.5}Mn_y/ZrO_2$ catalysts

The CO_2—TPD profiles for $Mg_{0.5}Mn_y/ZrO_2$ catalysts are depicted in Figure 4. Two peaks are observed in the CO_2—TPD profiles of $Mg_{0.5}/ZrO_2$ catalyst at 114 ℃ and 646 ℃ with 3∶1 in peak intensity, indicating that there are two different kinds of basic sites. Jiang et al. reported a very broad feature appeared between 50 ℃ and 400 ℃ in the CO_2—TPD profiles of MgO and MgO/ZrO_2 samples [19], similar to the low-temperature desorption peak in our CO_2—TPD profiles. However, the other desorption peak for the samples heated to the higher temperature was not studied in their work.

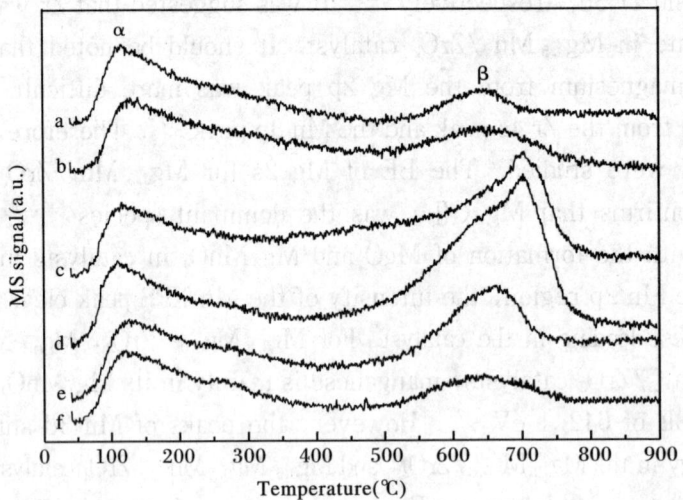

Figure 4 The CO_2—TPD features over (a) $Mg_{0.5}/ZrO_2$, (b) $Mg_{0.5}Mn_{0.05}/ZrO_2$, (c) $Mg_{0.5}Mn_{0.1}/ZrO_2$, (d) $Mg_{0.5}Mn_{0.2}/ZrO_2$, (e) $Mg_{0.5}Mn_{0.3}/ZrO_2$ and (f) $Mg_{0.5}Mn_{0.4}/ZrO_2$ catalysts.

The low-temperature desorption peak of the same peak intensity and desorption temperature was also obtained for Mn-modified catalyst. On the other hand, upon addition of a little Mn in the catalyst, the high-temperature desorption peak shifted to higher temperature and the intensity became stronger. With further increase of the Mn content in the catalysts, desorption temperature and intensity decreased. It is suggested that the $Mg_{0.5}Mn_y/ZrO_2$ catalyst has a larger share of sites with stronger basicity when certain amount of Mn is added.

3.5. Catalytic performance

The reaction of acetontrile with methanol to form acrylonitrile is a complicated reaction, and methanol often decomposes during the course of the reaction. Although a small amount ($< 1\%$) of methanol was consumed to form CH_4 and CO, the effect of the change in methanol concentration could be neglected in the reaction of acetonitrile with methanol over $Mg_{0.5}Mn_y/ZrO_2$ catalysts because of a fixed molar ratio of methanol-to-acetonitrile (10 : 1). The products containing nitrogen are acrylonitrile, propionitrile, and methacrylonitrile as confirmed by GC—MS analysis. Thus, the conversion and selectivity, calculated by equation (1) and (2), are reasonable.

3.5.1. Effect of Mg loading

Figure 5 shows the effects of Mg loading on the conversion and selectivity of the reaction of acetonitrile and methanol over $Mg_xMn_{0.2}/ZrO_2$ catalyst at 480 ℃. The catalysts investigated are $Mg_{0.4}Mn_{0.2}/ZrO_2$, $Mg_{0.5}Mn_{0.2}/ZrO_2$, $Mg_{0.6}Mn_{0.2}/ZrO_2$, $Mg_{0.7}Mn_{0.2}/ZrO_2$, and $Mg_{0.8}Mn_{0.2}/ZrO_2$. In order to reveal the effect of Mg loading, all other reaction conditions were fixed: 500mg catalyst, nitrogen flow rate at 70 mL/min, reactant (methanol + acetonitrile) flow rate at 0.03 mL/min, methanol/acetonitrile molar ratio at 10 : 1, and reaction temperature at 480 ℃. Under these conditions, the conversion and selectivity changed only slightly (about 16.2% and 84.0%, receptively)

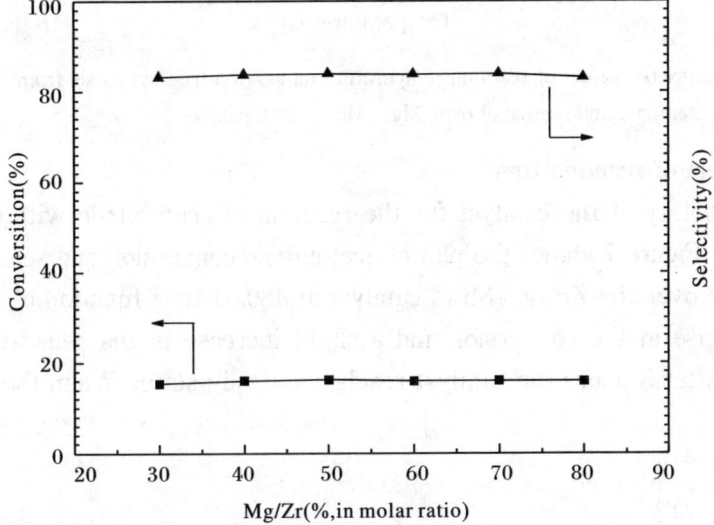

Figure 5 Results of acrylonitrile synthesis from acetonitrile and methanol over $Mg_xMn_{0.2}/ZrO_2$ catalysts with the different Mg loading.

3.5.2. Effect of temperature

Figure 6 shows the conversion and selectivity for the synthesis of acrylonitrile over a $Mg_{0.5}Mn_{0.2}/ZrO_2$ catalyst at the different reaction temperature. There were different optimum temperatures for the conversion of acetonitrile and selectivity to acrylonitrile. Acetonitrile conversion was the highest at 480 ℃, while the highest selectivity to acrylonitrile appeared at 400 ℃. Considering both acetonitrile conversion and selectivity to acrylonitrile, the optimum reaction temperature is 480 ℃. At this temperature, the yield of acrylonitrile is 13.6%.

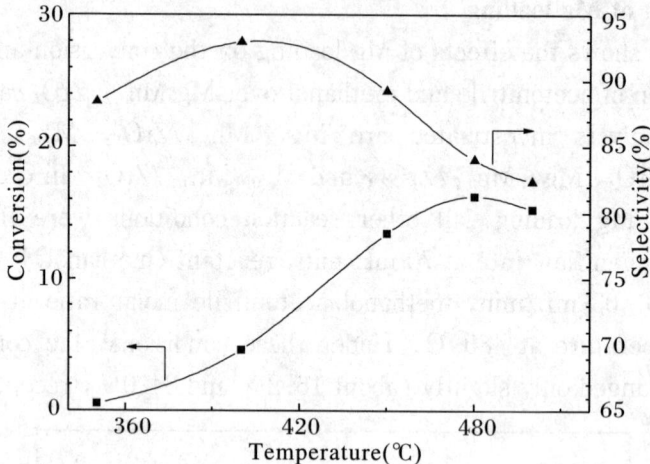

Figure 6 Effect of reaction temperature on acrylonitrile synthesis from acetonitrile and methanol over $Mg_{0.5}Mn_{0.2}/ZrO_2$ catalyst.

3.5.3. Effect of reaction time

The stability of the catalyst for the reaction of acetonitrile with methanol was tested. Figure 7 shows the plot of acetonitrile conversion and selectivity to acrylonitrile over the $ZrMg_{0.5}Mn_{0.2}$ catalyst at 480 ℃ as a function of time. An initial decrease in the conversion and a slight increase in the selectivity were observed. After 3 hours the catalyst reached a steady state. When the stability

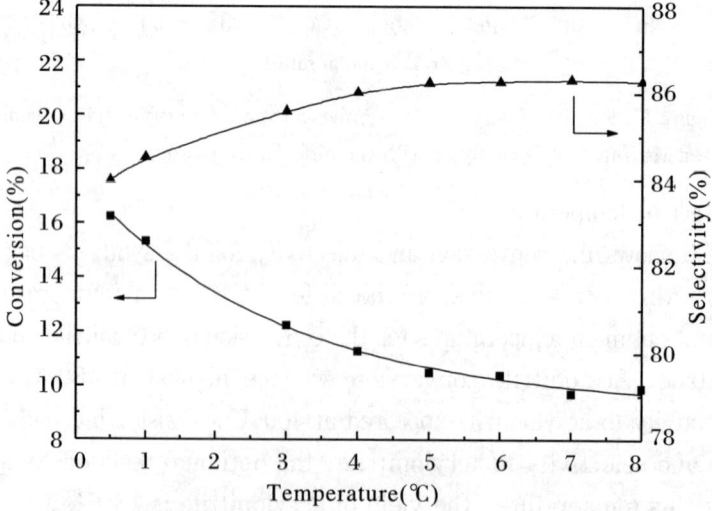

Figure 7 Activity change of $Mg_{0.5}Mn_{0.2}/ZrO_2$ catalyst for acetonitrile conversion as a function of time.

tests were conducted up to 8 h, it was found that the activity decreased very slowly with time. Coke formation on the surface seemed to be responsible for the decrease[7]; the catalyst became slightly darker than its initial pale yellow. To confirm this, an air oxidative reactivation was carried out after 8 h: calcinations of the used catalyst at 600 K for 2 h, followed by activation under nitrogen at the same temperature for 2 h. The regenerated catalyst showed almost the same activity as that of the fresh catalyst.

3.5.4. Promoting of the Mn-dopant

Figure 8 shows the results for the reaction of acetonitrile with methanol over $Mg_{0.5}Mn_y/ZrO_2$ catalysts at 480 ℃. The $Mg_{0.5}/ZrO_2$ catalyst is virtually inactive, and the conversion is only 1%. Similar phenomenon has also been reported for unsupported MgO catalyst in synthesis of acrylonitrile by methylenation of acetonitrile with methanol at 350℃ [7] and 400℃ [11]. The effects of Mn content on the activity are shown in Figure 8. One could see that the conversion first increased and then decreased with the increase of Mn content. The maximum conversion (~16.4%) obtained for $Mg_{0.5}Mn_{0.2}/ZrO_2$ catalyst, is higher than that obtained for unsupported Mn—MgO catalyst (9.6%) [7]. A slight decrease in conversion appeared with further increase of the Mn content. As shown in Figure 8, the addition of Mn in the catalysts also causes product distribution in the synthesis of acrylonitrile from acetonitrile and

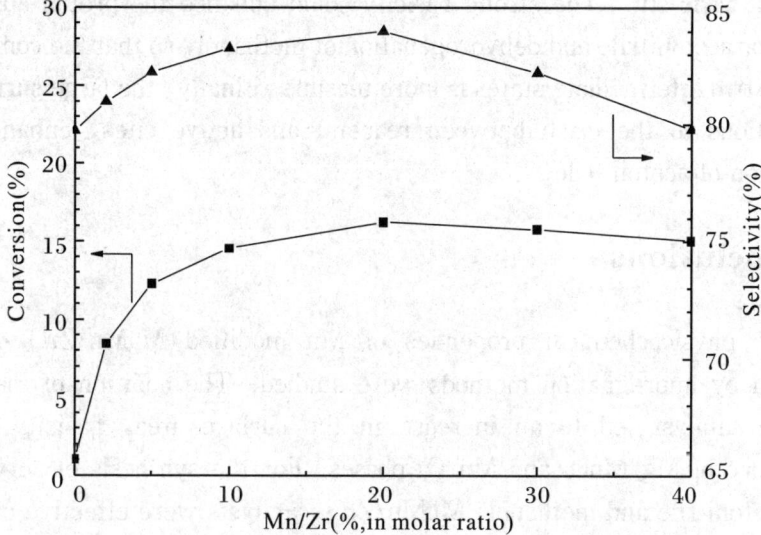

Figure. 8 Addition effect of Mn to $Mg_{0.5}/ZrO_2$ on the reaction of acetonitrile with methanol at 480 ℃.

methanol. Just like the conversion, the selectivity to acrylonitrile over the $Mg_{0.5}Mn_y/ZrO_2$ catalysts has a maximum value with the increase of the promoter content. At the optimum content of Mn-promoter ($Mn/Zr = 0.2$, in atomic ratio), the selectivity to acrylonitrile was 84.0%, slightly higher than that for $Mg_{0.5}/ZrO_2$ catalyst or 79.8% for $Mg_{0.5}Mn_{0.4}/ZrO_2$ catalyst (79.7%).

It can be seen from Figure. 8 that the activity and selectivity of $Mg_{0.5}Mn_y/ZrO_2$ catalyst is improved by addition of a suitable amount of Mn. Firstly, from the above characterizations, the higher catalytic activity of the $Mg_{0.5}Mn_{0.2}/ZrO_2$ catalyst could be partly attributed to the formation of Mg_2MnO_4 species. According to XPS and XRD results, the addition of Mn enhanced the conversion of MgO into Mg_2MnO_4, on which the reactant molecules could be more easily adsorbed than on MgO surface. However, further increase of Mn in the catalyst, especially at molar ratio of Mn/Mg in the catalyst higher than 1 : 2, resulted in the formation of the inactive Mn_2O_3 species. The Mn_2O_3 partially covered the active species, leading to low activity of catalyst in the synthesis of acrylonitrile from acetonitrile and methanol. As seen in Table 1, the surface areas of the $Mg_{0.5}Mn_y/ZrO_2$ catalyst increase with the increase in the Mn content. Secondly, from the CO_2—TPD profiles of the $Mg_{0.5}Mn_y/ZrO_2$ catalyst, it is seen that both the temperature and intensity of the high-temperature desorption peak for $Mg_{0.5}Mn_{0.2}/ZrO_2$ catalyst is the highest, corresponding to having the strongest basic site. The strong basicity could enhance the proton abstraction ability for acetonitrile and dehydrogenation of methanol, so that the combination of those two intermediate states is more feasible. Finally, the large surface area is propitious to the clash between reactant and active sites, enhancing the conversion of acetonitrile.

4. Conclusions

The physicochemical properties of Mn modified $MgMn/ZrO_2$ catalysts prepared by impregnation methods were studied. The addition of manganese into the catalyst led to an increase in the surface area, basicity and the appearance of Mg_2MnO_4 and Mn_2O_3 phases. For the synthesis of acrylonitrile from acetonitrile and methanol, $MgMn/ZrO_2$ catalysts were effective catalysts. An optimum concentration of manganese that maximized the generation of acrylonitrile existed and is associated with a maximum content of Mg_2MnO_4 and the basic sites. At the optimum Mn-content ($Mn/Zr = 0.2$, in atomic ratio), the conversion of acetonitrile is 16.2% and the selectivity to acrylonitrile is 84.0%. Simultaneously, the optimum conditions of MgMn/ZrO$_2$ catalysts for the reaction are that the Mg/Zr ratio is 50% in molar ratio and reaction temperature was 480 ℃.

Acknowledgments

This work is supported by Zhejiang Provincial Nature Science Foundation of China (No. RC00043).

References

[1] T. Ogawa, R. Cedeno, M. Inoue, Polym. Bull. 2 (1980) 275.
[2] J.R. Jennings, R.J. Cozens, K. Wade, Appl. Catal. A 130 (1995) 175.
[3] W.M. Zhang, P.G. Smirniotis, J. Catal. 182 (1999) 70.
[4] Y.C. Kim, W. Ueda, Y. Moro-oka, Catal. Today 13 (1992) 673.
[5] M.O. Guerrero-P rez, J.L.G. Fierro, M.A. Vicente, M.A. Ba ares, J. Catal. 206 (2002) 339.
[6] R.K. Grasselli, J.D. Burrington, J.F. Brazdil, Faraday Discuss. Chem. Soc. 72 (1982) 203.
[7] H. Kurokawa, T. Kato, W. Ueda, Y. Morikawa, Y. Moro-Oka, T. Ikawa, J. Catal. 126 (1990) 199.
[8] Y. Yamazaki, T. Kawai, Sekiyu Gakkai Shi 12 (1969) 693.
[9] J.M. Hur, B.Y. Coh, H.I. Lee, Catal. Today 63 (2000) 189.
[10] P.G. Smirniotis, W.M. Zhang, Appl. Catal. A 176 (1999) 63.
[11] L.H. Zhao, G.Y. Liu, Y.L. Xie, H.S. Zhou, L.M. Li, M.F. Luo, React. Kinet. Catal. Lett. 82 (2004) 219.
[12] W. Ueda, T. Yokoyama, Y. Moro-Oka, T. Ikawa, Ind. Eng. Chem. Prod. Res. Dev. 24 (1985) 340.
[13] C.L. Bothe-Almquist, R.P. Ettireddy, A. Bobst, P.G. Smirniotis, J. Catal. 192 (2000) 174.
[14] A. Trunschke, D.L. Hoang, J. Radnik, H. Lieske, J. Catal. 191 (2000) 456.
[15] PDFmaint Version 3.0, Powder diffraction database, Bruker Analytical X-ray systems GmbH, 1997.
[16] N.M. Deraz, Thermochimica Acta 421 (2004) 173.
[17] Handbook of X-ray Photoelectron Spectroscopy (Perkin-Elmer Corporation, 1992).
[18] D. Cáceres, I. Colera, I. Vergara, R. González, E. Román, Vacuum 67 (2002) 577.
[19] D.E. Jiang, G.C. Pan, B.Y. Zhao, G.P. Ran, Y.C. Xie, E.Z. Min, Appl. Catal. A 201 (2000) 169.

Appendices

附 录

文献导读部分参考译文
Translation of Literature Introduction
怎样写科技论文

1 什么是科技论文?

科技论文是一种以向读者阐明假说、数据和结论为目的的描述性文体。论文是研究工作的核心。如果你的研究没有写成论文,也就相当于你没有做研究。有意义却没有发表,等同于不存在。

要认识到你的研究的目的是为了形成并证实假说,从一些实验中得出结论,并将这些结论传授给别人,并不是简单地去"收集数据"。

一篇研究论文并不是一个简单地完成收集结果的程序,它也有助于形成继续研究的工作框架。如果你清楚写论文的目的,这可能有助于你开展研究工作。一个好的文章大纲同时也是一个好的研究计划,因此,在研究的过程中,你应当反复地修改这些计划(提要)。开始时,你应有完善的计划;结束时,应有充分总结。不断努力地去理解、分析、总结并最终在文章中形成假说。这是一个比较有效的方法,而不要等到收集完成后才去整理数据。

2 提纲

2.1 为什么要写提纲?

我重点强调提纲在论文写作、准备报告以及研究计划过程中的中心地位。我坚信对你我来说,最有效的方法是按照提纲进行写作。一个提纲就是一篇论文的写作计划,其中包括你所检测到的数据。事实上,你应当把提纲看做是一套具有详尽目的、假说和结论的数据,而不仅仅是列出各段的内容。

提纲本身应该包括简洁的文字。如果大家都认可提纲中的细节部分(即数据和结构),那么正文组织起来就相对容易。如果我们对提纲没有达成一致意见,那么写正文是没有意义的。写文章时,大部分时间用在写正文上;而大部分思考则是用在整理和分析数据中。相对有效的方法是在开始写之前,详细思考几遍写作提纲,写很多遍正文是很慢的。

所有我写的文章,包括论文、报告、建议(当然还有用于研讨会的幻灯片)都源自提纲。同样,我也鼓励你们学会使用它。

2.2 你应当怎样来组织提纲呢?

最经典的方法是找一张空白纸,用任何顺序,写下你所想到的与这篇文章有关的重要观点。问一下自己这些显而易见的问题:"为什么我要做这项工作?","它意味着什么?","我要验证什么样的假设?","我究竟验证了哪些假设?","结果是什么?","这项工作产生什么新方法或新物质? 是什么?","我都做了那些测试?","什么化合物? 它们是如何表征的?"。粗略写出相关的方程，画出图表和示意图。实际上也就是试图写出你的主要观点。如果你的研究开始是为证实某个假设，然而当你发现你的数据看上去可以更好地验证其他假设时，不要焦急。把它们两者都写出来，去选择假设、目的和数据的最佳组合。时常你会发现，当写完一篇论文时，最终得到的结果和开始时的目的是不同的。许多好的观点来自机遇和反复修正。

当你已经写下你所能写的，再拿出另外一页纸，试着草拟一份提纲。将你的观点分成三大类。

1. 引言

为什么我要做这项工作？主要的目的和假设是什么？

2. 结果和讨论

结果是什么？化合物是怎样合成的同时又是怎样表征的？测试方法是什么？

3. 结论

所有这一切意味着什么？什么样的假设被证实或否定？我得到了什么？为什么会得到不同的结果？

接下来，把每一部分再仔细组织。集中精力整理这些数据。尽可能以清晰、紧凑的图形、表格等把数据展现出来。或许这个过程漫长：我可能要用5~10次以不同的方式来构思一张图形，以便决定怎样做才最清楚（而且看上去最美观）。

最后，将所有这些——内容的提纲、表格、草图、方程式——排好顺序。

当你对你所收集的资料感到满意（或者你已经明确你还需要收集哪些额外的数据），并且有一个合理构架时，请将大纲交给我。简要地标明哪些地方还需要数据，你认为这些数据大概是怎样的，如果你的推测是正确的，你又会如何去解释它。拿到你的大纲后，我将把我的观点、建议反馈给你。一般，我们需要四或五个来回才能达成一致（中间经常还需要补做一些实验）。在我们的意见一致后，所有的数据通常以最终（或接近最终的）形式确定下来（也就是说，在提纲中的表格，图表等最终将成为文章中的表格，图表等）。

然后就可以开始写作了，而你写的这些大多将用于正文。

有效利用我们时间的关键是，我们应尽早地交换提纲和建议。在任何情况下，都不要等到你已经收集"全"了数据之后才开始动笔写提纲。研究是永无止境的。一旦你看到你的研究结果初具雏形时，就应该立刻开始构思文章和提纲，这将会节约大量的精力和时间。即便在认真组织成文前就已经决定有其他重要的实验需要补做，我们所写的提纲对我们的研究工作也是有帮助的。

2.3 提纲

一个提纲应该包括哪些内容呢？

1. 标题
2. 作者
3. 摘要

不要着急写摘要，可以等文章完成后再写。

4. 引言

文章开始的前1或2段应该完全用来写引言。特别注意要写好开头第一句话。最好是简洁地阐述工作目的，并指明该工作为什么重要。一般而言，引言应该包含以下几个要素：

- 工作目的。

- 对研究工作目的的合理性分析：该工作为什么很重要？

- 工作背景:别人做了什么工作？做得怎么样？我们以前做了哪些工作？
- 导读:读者应该注意该文章的哪些方面？有意义的要点有哪些？我们采用了哪些策略？
- 总结结论:读者期望什么样的结论？在提纲的前几个版本中,你应该包括实验部分中涉及的所有内容(在这一点上,就像是段落的副标题)。

5. 结果和讨论

通常,结论和讨论是合在一起的。这一部分应根据主题来进行组织。分段应有黑体字的副标题,目的是使文章组织更加清晰,帮助读者通览全文,找到他们感兴趣的内容。下面列举一些适合作副标题短语的例子:

- 烷基硫醇的合成
- 单层膜的表征
- 邻二醇单元的绝对构像
- 滞后现象与表面粗糙度的关系
- 温度对速率常数的影响
- 自交换速率随溶剂极化度而降低

尽可能使副标题具体并且内容丰富。例如,短语"The Rate of Self-Exchange Decreases with The Polarity of The Solvent"明显比"Measurement of Rates"长,但对读者更有帮助。一般来说,尽量概括该段落的共同点。

- 初始材料的合成
- 产物的表征
- 表征方法
- 测量方法
- 结果(速率常数,接触角,其他)

在提纲中,不要罗列大量的正文内容,而是要把数据放在合适的位置。任何正文应该简单地指明某段中包括了什么数据。

- 副标题
- 图表(附说明)
- 示意图(附说明和注解)
- 方程
- 表格(版式正确的)

记住要把文章看做是一个实验结果的集合,尽可能清晰和简洁地使用图表、表格、方程和示意图。论文中的正文是为解释数据服务的,因而它是次要的。可以被压缩进表格、方程等的信息越多,文章越短,越易读。

6. 结论

在提纲里,论文的结论可以概括成一些简短的短语或句子。除非是为了特殊强调之外,一般不要重复在结果部分已经有的结论。结论部分应该是像上面说的那样,而不仅仅只是一个总结。它应该增加一些新的、更高层次的分析,并且应该明确地指出这项工作的意义。实验部分包括所有实验部分的副标题,都要按照正确的顺序与结果相对应。

2.4 总结

- 在一个项目开始时,就应该着手去写可能的论文提纲,而不要等到研究结束的时候。研究可能永远没有结尾可言。

- 要围绕易于接受的数据、表格、方程式、图表、示意图来整理提纲和论文,而不是围绕文章而写文章。

- 要按重要性来整理,而不要按照时间顺序来整理。论文写作的一个重要细节是要权衡各部分的重要性。新手写作常按照时间顺序来写,即:他们常常从所珍爱的开始时的失败写起,直到最后的成功,照此顺序来叙述实验过程。这种方法是完全错误的。应该从最重要的结果写起,然后是较重要的结果。读者们通常不关心你是怎么得到的结果,而只关心结果是什么。短文章比长文章更易读。

3 一些英文文体上的要点

- 不要把名词误用为副词:

误:

ATP formation; reaction product

正:

formation of ATP; product of the reaction

- 单词"this"后面必须接名词,这样"this"的指代对象会更加清楚。

误:

This is a fast reaction; This leads us to conclude

正:

This reaction is fast; This observation leads us to conclude

- 描述实验结果一律要用过去时态。

误:

addition of water *gives* product

正:

addition of water *gave* product

- 尽可能使用主动语态。

误:

It was observed that the solution turned red.

正:

The solution turned red. 或

We observed that the solution turned red.

- 所有的比较都应该是完整的。

误:

The yield was higher using bromine.
正：
The yield was higher using bromine than chlorine.

• 文章排版时，要使用两倍行间距(不用一倍或一倍半)。在冒号、逗号和句末的句号后要空一格。要留出足够的页边空间(通常，在文章两侧、页首和页尾留出 1.25 英寸的空间)。

高比表面积 CuO—CeO$_2$ 催化剂中不同 CuO 物种的区分及这些 CuO 物种对 CO 氧化反应的催化活性贡献

本文采用改进的柠檬酸溶胶—凝胶法(在 N$_2$ 中预处理)制备了高比表面积(> 90 m$^2 \cdot$ g^{-1})的纳米 CuO—CeO$_2$ 催化剂。CO 程序升温还原实验(CO—TPR)结果表明，该 CuO—CeO$_2$ 催化剂中存在三种形式的 CuO 物种，分别为催化剂表面高分散的 CuO、晶相 CuO 和进入 CeO$_2$ 晶格的 Cu^{2+}。选取 CO 氧化反应作为模型反应，计算得到了每种 CuO 物种的催化活性。发现高分散的 CuO 物种具有最高的催化活性，晶相 CuO 其次，CeO$_2$ 晶格中的 Cu^{2+} 贡献最小。高分散的 CuO 物种具有最高的氧化活性(183.3 mmol$_{CO} \cdot$ g$_{Cu}^{-1} \cdot$ h^{-1})，晶相 CuO 其次(100.4 mmol$_{CO} \cdot$ g$_{Cu}^{-1} \cdot$ h^{-1})，CeO$_2$ 晶格中的氧化活性最小(21.3 mmol$_{CO} \cdot$ g$_{Cu}^{-1} \cdot$ h^{-1})。此外，在高温焙烧或者 CO 氧化反应的条件下，进入 CeO$_2$ 晶格的部分 Cu^{2+} 能够迁移到催化剂表面转变成高分散的 CuO。此外，通过催化剂经硝酸处理后不同温度焙烧后样品的 CO 氧化活性的变化和这些样品的 CO—TPR 图谱中还原峰的变化情况，结合 XPS 测得的催化剂硝酸处理前后表面 Cu 浓度的变化信息和样品的 CO 循环氧化实验，发现 CuO—CeO$_2$ 催化剂中三种 CuO 物种是可以相互转化的。高分散 CuO 经 800 ℃焙烧转化为晶相 CuO；而经历 600 ℃焙烧或者 CO 氧化反应后，进入 CeO$_2$ 晶格的部分 Cu^{2+} 可以转变成表面高分散的 CuO，从而可以增强催化活性。

关键词：CuO—CeO$_2$ 复合氧化物；溶胶—凝胶法；程序升温还原实验；CO 氧化反应

1 引言

近年来，CO 催化氧化反应由于大量应用在许多反应中如机动车尾气的控制、密闭系统中 CO 的消除、CO$_2$ 激光器中气体的净化、CO 气体传感器和燃料电池[1,2] 而引起人们的广泛关注。贵金属催化剂如金、铂、钯等在低温 CO 氧化反应方面具有良好的催化效果[1,3-5]。然而，由于贵金属催化剂高昂的价格和易中毒性，越来越多的注意力转向新的廉价的过渡金属催化剂。其中，Cu 催化剂被认为是一种良好的催化 CO 氧化的碱金属催化剂[6,7]。CuO—CeO$_2$ 催化剂由于具有高的催化活性和选择性而被广泛地用于各种催化反应中，如 NO 还原反应、CO 完全氧化反应、选择氧化反应、水煤气转换反应及 phenol 的湿法氧化反应[8-11]。Avgourpoulos 等[12] 报道了 CuO—CeO$_2$ 催化剂具有和 Pt/Al$_2$O$_3$ 催化剂近似的活性。Sedmak 等[8] 采用溶胶—凝胶法制备出了纳米结构的 Cu$_{0.1}$Ce$_{0.9}$O$_{2-y}$ 催化剂，这种催化剂在低温下对于富氢体系的 CO 选择氧化具有很高的催化活性和选择性。由于低廉的价格和较高的催化活性，CuO—CeO$_2$ 催化剂很有希望在将来取代贵金属催化剂[13]。

对于 CO 氧化反应来说,有两个因素制约着 CuO—CeO_2 催化剂的催化活性,分别是催化剂的比表面积和活性中心数目。尽管引起了人们的广泛关注,但是具有高比表面积($\geqslant 60 \text{ m}^2 \cdot \text{g}^{-1}$) 的 CuO—$CeO_2$ 催化剂几乎很少见报道。众所周知,具有高比表面积的催化剂因为能够提供更多的活性中心而带来更好的催化活性[14,15]。我们在前面的文章中[16]报道了一种用模板合成法制备的高比表面积纳米 CuO—CeO_2 催化剂。该催化剂在低温 CO 氧化和富氢体系的 CO 选择氧化方面显示出了高的催化活性。我们课题组在另外一篇文章中[17]报道了采用一种改进的溶胶-凝胶法制备出具有纳米微晶结构和高比表面积的 $Ce_{0.8}Pr_{0.2}O_Y$ 固溶体。同模板合成法相比,这种改进的溶胶—凝胶法具有制备方法简单,收率高的优点。

人们都相信 CO 氧化中 CuO—CeO_2 催化剂中的 CuO 物种是活性中心[18]。然而,CuO—CeO_2 催化剂中 CuO 的存在形式是比较复杂的,各种 CuO 对 CO 氧化的贡献更难区分。在我们以前的研究[16]中发现用模板合成法制备的 CuO—CeO_2 催化剂中高分散 CuO 是 CO 催化氧化的活性组分,但是没有涉及其他形式的 CuO 物种对 CO 催化氧化的贡献。

为了更进一步研究这个问题,在本文中,我们采用改进的柠檬酸溶胶—凝胶法制备出了具有高比表面积的 CuO—CeO_2 催化剂,这些催化剂对 CO 氧化反应显示出了高的催化活性。对催化剂体系中的 CuO 物种进行了区分并计算了各种组分对 CO 氧化的贡献,并且研究了在反应中这些 CuO 物种的转化情况。

2　实验部分

2.1　催化剂的制备

采用两种不同的方法来制备催化剂:分别称为传统的柠檬酸溶胶—凝胶法和改进的柠檬酸溶胶—凝胶法。

采用改进的柠檬酸溶胶—凝胶法制备不同 CuO 含量(5 mol %, 10 mol %, 20 mol %, 50 mol %) 的 CuO—CeO_2 催化剂,参见文献[19]。按照一定的化学计量比称取适量的 $Cu(NO_3)_2 \cdot 3H_2O$ 和 $Ce(NO_3)_3 \cdot 6H_2O$,用去离子水溶解,加入 2 倍于金属阳离子摩尔量的柠檬酸得到柠檬酸盐溶液,然后溶液用水浴加热形成凝胶。在这个过程中,溶液的颜色由蓝变绿。凝胶于 110 ℃ 烘干过夜,得到蓬松的物质即催化剂前驱体。再将前驱体置于管式炉中,在氮气中 800 ℃ 焙烧 2 h,使得柠檬酸在无氧的条件下分解碳化,得到黑色含有碳粒和氧化物的混合物(中间混合物)。接下来,得到的中间体在空气中 400 ℃ 焙烧 4 h 以烧掉碳粒。CuO 的实际含量是用原子吸收分光光度计(AAS)测定的。命名上,N8A4-5 催化剂表示对应的催化剂是先在 N_2 中 800 ℃ 焙烧 2 h,然后在空气中 400 ℃ 焙烧 4 h,而催化剂中 Cu 的含量是 5 mol %。

相应的,根据传统的柠檬酸溶胶—凝胶法[20]制备出一系列 CuO 含量为 10 mol % 的 CuO-CeO_2 催化剂。将前驱体直接在空气中 400 ℃ 或者 800 ℃ 焙烧 4 h,样品分别标记为A4-10或A8-10。

对于硝酸处理催化剂,取 1 g N8A4-10 催化剂,用 15 mL 50% 的稀硝酸浸泡 2 h,然后过滤,用大量蒸馏水冲洗以去除残留的硝酸和其他杂质,然后分别在 100 ℃、200 ℃ 烘干或在 600 ℃ 焙烧,依次标记为 N8A4-10H1、N8A4-10H2 和 N8A4-10H6。用同样的方式,A8-10 催化剂经硝酸处理后 100 ℃ 烘干制得样品 A8-10H1。

2.2　催化剂的表征

X 射线粉末衍射(XRD)分析在荷兰 Philips 公司的 PW 3040/60 型全自动 X 射线衍射仪上进

行。Cu Kα 射线,管电压 40 kV,管电流 40 mA。在 25 ℃ 条件下扫描范围为 20°~130°,扫描速度为 1.2°·min^{-1}。催化剂的微观参数应用 MAUD 软件通过 Rietveld 方法[21]采用 MAUD 软件[22]计算得到。采用 CeO_2 在空气中 1450 ℃ 焙烧 10 h 得到的样品作为标准样品来校正仪器微观结构参数。

催化剂的比表面积采用英国 Quantachrome Autosorb-1 型 N_2 物理吸附仪测定,77 K,N_2 吸附。表面积的计算采用 BET 公式。

透射电镜(TEM)实验是在 JEM1200 EX 电子显微镜上进行的,扫描电压 80 kV。

XPS 实验在美国 PHI 公司的 PHI 5000C ESCA System(经过美国 RBD 公司升级)上进行;采用条件为铝/镁靶(1253.6 eV)。并采用 AugerScan 3.21 软件进行数据分析。以 C1s = 284.6 eV 为基准进行结合能校正。采用 AugerScan 3.21 或 XPS Peak 4.1 软件进行分峰拟合。

$CuO—CeO_2$ 催化剂的还原性能采用 CO—程序升温还原实验(CO—TPR)来测定。将 50 mg 试样置于连接着自制 TPR 装置的石英管中,以 20 ℃·min^{-1} 的升温速率从 30 ℃ 至 600 ℃ 进行程序升温。

通入 CO 气体,流速为 30 mL·min^{-1},对样品以 20 ℃·min^{-1} 的升温速率进行程序升温。用质谱在线检测升温过程中的 CO_2(m/e = 44)和 CO(m/e = 28)信号的变化。反应混合气的组成为 5% CO、95% Ar。生成的 CO_2 气体的信号采用瑞士 Balzers 仪器公司生产的 Omnstar 200 型质谱仪在线分析,m/e = 44。

2.3 反应活性评价

以 CO 氧化反应的催化活性是在固定床(内径 6 mm)装置上进行评价的。装入为 250 mg 催化剂(20—40 目)。反应原料气的组成为(体积):1% CO,1% O_2 和 98% N_2。反应的气流量为 40 mL·min^{-1},对应的空速为 9600 mL·g^{-1}·h^{-1}。催化剂直接置于反应气中在反应温度下稳定一段时间,不需进行预处理。反应温度由置于催化剂中间部位的热电偶测定。反应体系中 CO 混合气的浓度均由装有 HP PLOT column (30 m × 0.32 mm × 12.0 μm)柱子的 Agilent 6850 气相色谱仪检测,检测器为 TCD 检测器。

3 结果

3.1 结构表征

表 1 列出了采用两种制备方法制得的 CuO 含量为 10 mol % 的 $CuO—CeO_2$ 催化剂的比表面积。样品 N8A4-10 的比表面积高达 131 m^2·g^{-1},远远高于 A4-10 的比表面积(60 m^2·g^{-1})和而 A8-10 的比表面积(10 m^2·g^{-1})。样品 A8-10 的比表面积急剧下降的原因可能是高温焙烧引起催化剂的晶粒烧结、长大[23]。比较这几种催化剂的比表面积,可以看出经过 N_2 预处理的改进的柠檬酸溶胶—凝胶法可以制备出具有高比表面积的催化材料。

不同 CuO 含量的 $CuO—CeO_2$ 催化剂的比表面积同样列于表 1。从表中看到当加入 CuO 含量低于 10 mol % 时,催化剂的比表面积随着 CuO 含量的增加而增大,预示着 Cu 的加入会增大催化剂的比表面积。然而,当 CuO 含量从 10 mol % 到 50 mol % 时,催化剂的比表面积反而下降,这可能是当 CuO 含量超过 10 mol % 时有晶相 CuO 生成的缘故。

图 1 给出了 $CuO—CeO_2$ 催化剂的 XRD 图谱。CuO 含量为 10 mol % 的 $CuO—CeO_2$ 催化剂的 XRD 图谱见图 1a,从图中可以看出所有样品都出现 CeO_2 的特征衍射峰。样品 A8-10 在 35.6°和 38.8°处出现了微弱的 CuO 特征峰,说明催化剂中形成了晶相 CuO。图 1b 是不同 CuO 含量的

CuO—CeO₂ 催化剂的 XRD 图谱。同样，所有样品都能观测到 CeO₂ 的特征衍射峰。然而，当 CuO 含量低于 20 mol % 时没有观察到晶相 CuO 的特征峰；当 CuO 含量达到 20 mol % 时，才出现了微弱的 CuO 特征衍射峰，随着 CuO 含量进一步提高至 50 mol %，这些衍射峰的强度明显增强。

由 XRD 仪器检测得到的 CuO—CeO₂ 催化剂的相组成、晶胞参数和平均晶粒大小同样列于表 1。从表中可以看出所有样品都含有立方相 CeO₂ 结构。CuO 含量较高的样品（N8A4-20 和 N8A4-50）以及经历高温焙烧的样品（A8-10）出现了 CuO 结构。

很明显，随着 CuO 含量增加到 20 mol % 时，催化剂的晶胞参数是逐渐减小的，当 CuO 含量在 20 mol % 和 50 mol % 之间，晶胞参数保持不变。

样品 N8A4-10、A4-10 和 A8-10 的平均晶粒大小分别为 7.3 nm、8.2 nm 和 63.6 nm。对于不同 CuO 含量的 CuO—CeO₂ 催化剂来说，纯的 CeO₂ 平均晶粒最大（10.9 nm），其他 CuO—CeO₂ 催化剂的晶粒在 6.0 nm 至 8.0 nm 之间。催化剂不同方向上的平均晶粒大小的差异说明存在各向异性。

图 2 为样品 N8A4-10、A4-10 和 A8-10 的 TEM 图。从图 2a 中可以看出，样品 N8A4-10 晶粒大小接近于 7 nm，而样品 A4-10 的晶粒大小在 9 nm 左右。与之对应的是，样品 A8 具有最大的晶粒，在 60 nm 左右。这些结果和由 XRD 结果根据谢乐公式计算得到的相吻合。

3.2 催化活性测试

图 3 给出了不同制备方法制得的 CuO—CeO₂ 催化剂的 CO 氧化活性曲线。CuO 含量为 10 mol % 的 CuO—CeO₂ 催化剂的 CO 氧化活性结果见图 3a。从图中可见，N8A4-10、A4-10 和 A8-10 的 T_{90}（转化率为 90% 时所需的反应温度）分别为 100 ℃、110 ℃ 和 120 ℃。由改进方法制得的一系列 CuO—CeO₂ 催化剂的 CO 氧化活性结果见图 3b。纯的 CuO 和 CeO₂ 样品的 CO 氧化活性很差。当 CuO 含量低于 10 mol % 时，随着 CuO 的掺杂，催化剂的 CO 氧化活性明显提高；当继续增加 CuO 的含量时 CuO—CeO₂ 催化剂的 CO 氧化活性基本不变。

3.3 CO—程序升温还原实验（CO—TPR）

采用 CO—TPR 实验来表征催化剂的还原能力。图 4a 是催化剂 N8A4-10、A4-10 和 A8-10 的 CO—TPR 图谱。从图中可以看出，催化剂 N8A4-10 和 A4-10 在 200 ℃ 都有两个低温还原峰：位于 120 ℃ 附近的 α 峰和 180 ℃ 附近的 β 峰。同 A4-10 相比，N8A4-10 的 α 峰面积较大。而催化剂 A8-10 只有位于 180 ℃ 附近的 β 峰和在 β 峰高温方向的肩峰 γ。

由改进方法制得的一系列不同 CuO 含量的 CuO—CeO₂ 催化剂的 CO—TPR 图谱见图 4b。众所周知，纯的 CeO₂ 在 430 ℃ 和 900 ℃ 附近有 2 个还原峰，分别归属为是 CeO₂ 表面氧和晶格氧的还原[20]。同时，纯的 CuO 大约在 280 ℃ 左右有个 CO 还原峰[24]。当掺杂 CuO 后，催化剂 N8A4-5 在 200 ℃ 以下出现 2 个 CO_2 的形成峰：120 ℃ 附近的微弱的峰 α，和 175 ℃ 附近的信号较强的峰 β。随着 CuO 含量从 5 mol % 增至 10 mol %，α 和 β 峰的强度增大；然而，当 CuO 含量从 10 mol % 增至 50 mol % 时，α 峰没有明显的变化，β 峰继续宽化，面积进一步变大，同时在 195 ℃ 附近出现了一个新的肩峰（γ 峰）。

4 讨论

4.1 CuO—CeO₂ 催化剂中 Cu 物种的区分

如前面（图 1）提到的，当样品中 CuO 含量较低时观测不到 CuO 的特征衍射峰，这可能是由于固溶体的形成[11]或者 CuO 可能是以高分散的形式存在于 CeO_2 表面,这部分 CuO 的晶粒很小,以至于不能被 XRD 检测到[24]。当 CuO 含量达到 20 mol % 时 XRD 能检测到 CuO 的存在,这说明催化剂表面生成了晶相 CuO。同时发现采用改进的柠檬酸溶胶—凝胶法制得的催化剂晶粒小于采用传统方法制得的催化剂,这同 XRD (表1)结果和 TEM (图2)结果一致。这可能是因为采用改进的柠檬酸溶胶—凝胶法,前驱体在空气中焙烧前先在 N_2 中高温预处理,柠檬酸没有完全燃烧而是生成碳粒,碳粒能把铜铈氧化物相互隔离,阻止烧结。此外,CuO 的掺杂抑制了 CeO_2 晶粒的长大,而 $CuO—CeO_2$ 催化剂的晶粒同加入的 CuO 的没有明显的关系,这同 Fu 等[5]的研究结果一致。随着 CuO 含量的增加,催化剂的晶胞参数逐渐减小,这是由于 Cu^{2+} 的半径(0.072 nm)小于 Ce^{4+} 的半径(0.097 nm),当部分 Cu^{2+} 取代 Ce^{4+} 进入 CeO_2 晶格后导致晶格收缩。这表明这些催化剂都形成了 $Cu_xCe_{1-x}O_{2-\delta}$ 固溶体[25]。

α 峰是催化剂表面高分散的 CuO 的还原,β 峰是进入 CeO_2 晶格的 Cu^{2+} 的还原,γ 峰归属于晶相 CuO 的还原。所以催化剂 N8A4-10 和 A4-10 含有高分散的 CuO 和进入 CeO_2 晶格的 Cu^{2+},而催化剂 A8-10 含有进入 CeO_2 晶格的 Cu^{2+} 和晶相 CuO。催化剂 N8A4-10 的 α 峰面积比 A4-10 的大,说明催化剂 N8A4-10 含有更多的高分散 CuO 物种。而高分散的 CuO 是 CO 氧化反应的活性中心[41,70],所以催化剂 N8A4-10 的 CO 氧化活性优于 A4-10(见图1)。同时从图2b 中可以看到,催化剂 N8A4-20、N8A4-50 的 α 峰面积同 N8A4-10 的近似一致,因此它们的 CO 氧化活性也近似一致(见图1)。这也进一步验证了高分散的 CuO 是 CO 氧化反应的活性中心。然而,虽然没有 α 峰,样品 A8-10 仍然表现出一定的 CO 氧化活性,尽管其活性较差。而催化剂 A8-10 中主要存在进入 CeO_2 晶格的 Cu^{2+} 和晶相 CuO,说明其他 CuO 物种对 CO 反应也有一定的催化作用。

CO—TPR 结果(图4)表明 $CuO—CeO_2$ 催化剂上都能观测到 α 和 β 还原峰,而另外一个还原峰 α 则只有样品 A8-10、N8A4-20 和 N8A4-50 才出现。文献对这些低温还原峰有不同的解释。Avgouropoulos 和 Ioannides[26]认为 α 峰是与 CeO_2 发生强相互作用的 CuO 物种产生的,β 峰是与 CeO_2 作用力较弱的 CuO 物种的还原;Zou 等[27]则认为 α 峰归属于簇状 CuO 物种的还原,β 峰是孤立的 Cu^{2+} 的还原。Luo 等[18]也报道了两种还原峰,他认为 α 峰由表面高分散的 CuO 的还原引起,而 β 峰是晶相 CuO 的还原。Shiau 等有类似的报道[28,29]。本文中,从 XRD 结果我们发现所有 $CuO—CeO_2$ 催化剂均形成了 $Ce_xCu_{1-x}O_{2-\delta}$ 固溶体,而在 Cu 含量较高的样品中有晶相 CuO 的存在。我们知道,高分散 CuO 是很容易被还原的。因此,我们认为 α 峰是催化剂表面高分散的 CuO 的还原;β 峰是进入 $Ce_xCu_{1-x}O_{2-\delta}$ 固溶体的 Cu^{2+} 的还原。γ 峰归属于晶相 CuO 的还原,因为样品 A8-10、N8A4-20 和 N8A4-50 的 XRD 结果显示含有晶相 CuO。因此可以得出如下结论:$CuO—CeO_2$ 催化剂中主要存在三种形式的 CuO 物种,分别是催化剂表面高分散的 CuO,进入 CeO_2 晶格的 Cu^{2+} 和表面的晶相 CuO。在我们前面的工作[16]中有过类似的论述。

4.2 不同 CuO 物种对氧化活性的贡献

本文报道的 $CuO—CeO_2$ 催化剂比其他文献[15,30]报道的 $CuO—CeO_2$ 催化剂有更高的低温 CO 氧化活性,说明这种改进的柠檬酸溶胶—凝胶法可制得高活性的催化剂。在我们以前的研究[16,18]中报道了高分散 CuO 是 CO 催化氧化的活性中心,这在本文中也得到了证实。从图3中可以看出,催化剂 N8A4-10 比 A4-10 和 A8-10 的 CO 氧化活性高,而催化剂 N8A4-10、N8A4-20 和 N8A4-50 的 CO 活性基本一致。

催化剂的 CO—TPR 结果(图4)显示,样品 N8A4-10 的 α 峰面积比 A4-10 和 A8-10 的大。催化剂 N8A4-20、N8A4-50 的 α 峰面积同 N8A4-10 的近似一致,说明 CO 氧化活性同高分散的 CuO 物种有关。然而,虽然没有 α 峰,样品 A8-10 仍然表现出一定的 CO 氧化活性,尽管其活性较差。

说明其他 CuO 物种对 CO 反应也有一定的催化作用。

为了阐述三种 CuO 物种对 CO 氧化的贡献,我们做了样品 N8A4-10 和 A8-10 的酸处理实验。众所周知,表面高分散的 CuO 和晶相 CuO 很容易被硝酸溶解[18]。通过原子吸收分析,样品 N8A4-10H1 和样品 A8-10H1 的实际 CuO 含量分别为 4.95 mol % 和 4.96 mol %。催化剂 N8A4-10 和 A8-10 经硝酸处理前后的活性变化见图5。从图中可以明显地看出,样品 N8A4-10 和 A8-10 经硝酸处理后活性均有显著的下降。然而我们注意到经硝酸处理后的样品 N8A4-10H1 和 A8-10H1 活性基本一样。为了解释这个结果,我们对上述样品做了 CO—TPR 表征,结果见图6。正如前面所述,样品 N8A4-10 的 CO—TPR 图谱出现两个低温还原峰(α 和 β),然而经硝酸处理后,α 峰消失,仅剩下 β 峰。同样品 A8-10 相比,A8-10H1 的 γ 峰消失,β 峰变得更加对称。因此,催化剂的活性来自 CeO_2 晶格中的 Cu^{2+}。同样,样品 N8A4-10H1 和 A8-10H1 的 β 峰面积一致可以解释为什么这两种样品的活性基本一致。这些结果进一步证实了 α 峰是由催化剂表面高分散的 CuO 物种还原引起的,γ 峰是由晶相 CuO 的还原引起的,而 β 峰则归属于进入 CeO_2 晶格的 Cu^{2+} 的还原。

根据 100 ℃ 时 CO 氧化反应的转化率,我们通过计算得到了以 Cu 为单位计算的 $CuO—CeO_2$ 催化剂的单位反应速率,结果列于表2。由于在 100 ℃ 时具有不同的 CO 转化率,这几个样品显示出了不同的单位反应速率。其中样品 N8A4-10 的最高(为 101.8 $mmol_{CO} \cdot g_{Cu}^{-1} \cdot h^{-1}$),样品 A8-10 的居中 (60.6 $mmol_{CO} \cdot g_{Cu}^{-1} \cdot h^{-1}$),而 N8A4-10H1 和 A8-10H1 的最小(21.3 $mmol_{CO} \cdot g_{Cu}^{-1} \cdot h^{-1}$)。考虑到样品 N8A4-10 含有高分散的 CuO 物种和进入 CeO_2 晶格的 Cu^{2+} 物种,样品 A8-10 含有晶相 CuO 物种和进入 CeO_2 晶格的 Cu^{2+} 物种,而 N8A4-10H1 和 A8-10H1 仅含有进入 CeO_2 晶格的 Cu^{2+} 物种,我们可以把这三种 CuO 物种各自的单位反应速率计算出来,结果同样列于表2。从表中可以明显地看到:高分散的 CuO 物种对 CO 氧化贡献最大(183.3 $mmol_{CO} \cdot g_{Cu}^{-1} \cdot h^{-1}$),晶相 CuO 其次 (100.4 $mmol_{CO} \cdot g_{Cu}^{-1} \cdot h^{-1}$),$CeO_2$ 晶格中的 Cu^{2+} 贡献最小(21.3 $mmol_{CO} \cdot g_{Cu}^{-1} \cdot h^{-1}$)。

4.3 Cu 物种从晶格到表面的转变

图7给出了催化剂 N8A4-10 经硝酸处理后不同温度焙烧的 CO 氧化活性。我们发现 N8A4-10 经硝酸处理后的样品的 CO 氧化活性跟焙烧温度有关。随着样品 N8A4-10H1 焙烧温度的提高,样品的 CO 氧化活性也显著提高。

上述现象可以通过它们的 CO—TPR 图谱(见图8)来解释。从图中可以明显地看到,而经硝酸处理后的样品 N8A4-10H1 和 N8A4-10H2 的 α 峰消失,仅剩下 β 峰。一个有趣的现象是当样品 N8A4-10 经硝酸处理后 600 ℃ 焙烧(即 N8A4-10H6),由催化剂表面高分散的 CuO 物种还原所引起的 α 峰重新出现,同时归属于 CeO_2 晶格的 Cu^{2+} 物种还原所引起的 β 峰面积减小。然而总的峰面积没有变化,说明可还原的 CuO 物种总量不变。因此,样品 N8A4-10H6 中高分散的 CuO 物种的出现应该同 CeO_2 晶格的 Cu^{2+} 物种的减少有关。也就是说,当硝酸处理后的样品 600 ℃ 焙烧后,进入 CeO_2 晶格的部分 Cu^{2+} 转变成 $CuO—CeO_2$ 催化剂表面高分散的 CuO。

为了更好地了解催化剂在硝酸处理前后表面组成的变化,我们进行了 XPS 表征。表3中列出了催化剂 N8A4-10 经硝酸处理前后表面 Cu 浓度的变化。从表中可以看出,对于未经硝酸处理的样品 N8A4-10,表面 Cu^{2+} 的浓度(15.26 mol %)远远高于内部 Cu^{2+} 的浓度(9.58 mol %),这是催化剂中 CuO 物种在表面富集的缘故。然而经硝酸处理后,样品 N8A4-10H1 表面 Cu^{2+} 的浓度(5.73 mol %)接近于内部 Cu^{2+} 的浓度(4.95 mol %),这说明经硝酸处理后,催化剂表面的 CuO 已基本被除去。另外,样品 N8A4-10H6 表面 Cu^{2+} 的浓度达到 11.17 mol %,比内部 Cu^{2+} 的浓度(4.95 mol %)高得多。这些结果显示催化剂中 CuO 物种在表面再次富集,这同 CO—TPR

结果一致,并且充分说明催化剂表面的 CuO 对 CO 氧化起主要作用,CO 氧化活性的变化主要是由于高分散 CuO 在表面富集的缘故。

为了进一步证明上述结论,做了样品 N8A4-10H1 的 CO 循环氧化实验。在这个实验过程中,反应温度从 80 ℃ 升高到 220 ℃(相应的转化率从 10 % 左右升高至 100%),然后再从 220 ℃ 降至 80 ℃,整个过程称作一个循环[31]。样品在整个循环氧化实验过程中的催化活性见图 9。由图可见,在第一个循环的升温阶段样品 N8A4-10H1 的活性较差,随着循环实验的继续进行其活性逐渐提高。四个循环反应之后,样品 N8A4-10H1 的催化活性接近于样品 N8A4-10H6 的活性。图 10 对样品 N8A4-10H6 和 N8A4-10H1 经历四个循环反应之后的 CO—TPR 图谱进行比较。结果发现样品 N8A4-10H1 经历四个循环反应之后,归属于催化剂表面高分散的 CuO 物种的还原峰 α 再次出现,峰顶温度在 130 ℃,同 N8A4-10H6 的图谱类似。这些结果说明经过 CO 循环反应,CeO_2 晶格中的部分 Cu^{2+} 同样可以转变成 $CuO—CeO_2$ 催化剂表面高分散的 CuO。

示意图 1 给出了 $CuO—CeO_2$ 催化剂中三种 CuO 物种的转化示意图。如图所示,高分散的 CuO 经 800 ℃ 高温焙烧可以转化为晶相 CuO。这两种 CuO 都可以被硝酸溶解而除去。$CuO—CeO_2$ 催化剂经硝酸处理后仅剩下 CeO_2 晶格中的 Cu^{2+},而当催化剂经硝酸处理后再经历 600 ℃ 焙烧或者 CO 氧化反应时,部分 CeO_2 晶格中的 Cu^{2+} 可以从晶格中迁移到表面而形成高分散的 CuO。

5 结论

采用改进的柠檬酸溶胶—凝胶法制得了高比表面积的纳米 $CuO—CeO_2$ 催化剂。CO—程序升温还原实验(CO—TPR)结果表明,该 $CuO—CeO_2$ 催化剂中存在三种形式的 CuO 物种,分别为催化剂表面高分散的 CuO、晶相 CuO 和进入 CeO_2 晶格的 Cu^{2+}。增加高温 N_2 处理,能够有效地阻止催化剂的晶粒长大,获得具有高比表面积的催化材料。高分散的 CuO 物种对 CO 氧化活性贡献最大,CeO_2 晶格的 Cu^{2+} 活性贡献最小。硝酸处理能够洗去催化剂中的自由 CuO 物种(高分散和晶相 CuO),导致催化剂催化活性的显著下降。当酸洗过的样品经历高温焙烧或者 CO 氧化反应时,催化活性会得到部分恢复,因为部分 CeO_2 晶格中的 Cu^{2+} 可以从晶格中迁移到表面而形成高分散的 CuO。

N-甲基-2-吡咯烷酮与 3 种国产烟煤中某些有机物强烈缔合的证据

用二硫化碳/N-甲基-2-吡咯烷酮(CS_2/NMP)混合溶剂(体积比 1∶1)在室温下通过超声辐射分别萃取了采自神府、黑岱沟和肥城煤田的 3 种国产烟煤,常压蒸馏除去 CS_2 并在 110 ℃ 下减压蒸馏除去大部分 NMP 后得到含少量 NMP 的混合溶剂萃取物(MSEFs)。用丙酮在室温下通过超声辐射和索氏萃取分离各 MSEF,得到 MSEF 的丙酮萃取物(AEF1)。用丙酮直接萃取烟煤,得到烟煤的丙酮萃取物(AEF2)。GC/MS 分析表明,AEF1 中一系列成分的质谱基峰和次峰的 m/z 为 98,而在 AEF2 中没有检测出这些成分。NMP 本身的质谱基峰 m/z 为 99,m/z 为 98 的质谱基峰和次峰应该是 NMP 失去 1 个 α—H 形成的,即 NMP 与 3 种烟煤中的某些有机物

(OSs)发生了强烈的缔合作用,所检测出的质谱基峰和次峰的 m/z 为 98 的成分应该是 NMP-OS 缔合体。

关键词：N-甲基-2-吡咯烷酮；烟煤；缔合；GC/MS 分析

饭野雅等发现二硫化碳/N-甲基-2-吡咯烷酮(CS_2/NMP)混合溶剂(体积比 1∶1)在室温下对一些烟煤具有优良的溶解能力[1,2],该发现受到煤化学研究者的高度关注[3-6]。Zong 等对 CS_2 与 NMP 的热反应的研究表明,CS_2 中的 C==S 键与 NMP 中的 C==O 键之间具有强烈 π-π 相互作用,推测所形成的缔合体与煤中有机质之间的缔合作用是该混合溶剂使煤增溶的重要原因[7,8]。量子化学计算的结果也表明 CS_2 与 NMP 之间存在 π-π 相互作用[9,10]。但是,迄今尚未见有关煤中有机质与 CS_2—NMP 缔合体或与 NMP 之间发生强烈缔合作用的直接证据的报道。

我们分别用丙酮直接萃取了 3 种国产烟煤及其 CS_2/NMP 混合溶剂的萃取物(mixed solvent-extractable fraction,简称 MSEF),通过 GC/MS 分析比较了所得丙酮萃取物(acetone-extractable fraction,简称 AEF)组成的差别,发现 MSEF 的 AEF(简称 AEF1)中含有多种质谱碎片基峰或次峰 m/z 为 98 的成分,而用丙酮直接萃取煤样所得 AEF(简称 AEF2)中没有这些成分。本文讨论了该质谱碎片离子(fragmental ion,简称 FI)的来源,指出该 FI 提供了 NMP 与这些烟煤中某些有机物(OSs)强烈缔合的证据。

1 实验

1.1 主要仪器设备

用于煤样萃取和萃取液浓缩的设备包括上海医用仪器厂产的 CQ50 型超声波清洗器、日立公司产的 Hitachi himac CR 22E 型离心机、江苏省建湖县明星玻璃仪器厂加工的索氏萃取器和 Buchi 公司产的 Buchi R-134 型旋转蒸发器,用于萃取液分析的仪器是惠普公司产的 Hewlett Packard 6890/5973 型 GC/MS。

1.2 煤样和试剂

所用煤样分别采自陕西神府、内蒙古黑岱沟和山东肥城煤田,分别简称 SFC、HDGC 和 FCC。破碎和研磨后过 200 目筛(粒径< 75 μm),在 80 ℃下真空干燥 24 h 并在保持真空的状态下冷却至室温后取出,保存在干燥器中备用。煤样的工业分析和元素分析结果汇总于表 1 中。所用 CS_2、NMP 和丙酮均为分析纯试剂。

1.3 煤样的萃取和萃取液的 GC/MS 分析

将称取的 5 g 煤样及依次量取的 250 mL NMP 和 250 mL CS_2 加入到 1000 mL 的锥形瓶中,在超声辐射下萃取 2 h 后转移萃取液和萃余物的混合物至离心管中在 10000 rpm 下离心分离 10 min。用孔径为 0.8 μm 的聚四氟乙烯滤膜过滤离心管中的萃取液后以同量的 CS_2/NMP 混合溶剂萃取萃余物。以上操作重复 12 次。合并各次滤液,用旋转蒸发器在常压蒸馏除去 CS_2 后在 110 ℃下减压蒸馏除去绝大部分 NMP,得到含有少量 NMP 的 MSEF。在超声辐射下每次用 300 mL 丙酮多次清洗煤样的 CS_2/NMP 混合溶剂萃取物,直至过滤后的清洗液中不含 NMP。在 80 ℃下真空干燥清洗后的萃余物 24 h 并在保持真空的状态下冷却至室温后取出称重,根据干燥萃余物的干基重量 $W_{R,d}$ 利用减差法计算各由煤样所得 MSEF 的收率 Y：

$$Y = (W_d - W_{R,d})/W_{daf}$$

式中 W_d 和 W_{daf} 分别表示干基和有机基煤样重量。

在超声辐射下用 300 mL 丙酮萃取 MSEF 2 h 后用孔径为 0.8 μm 的聚四氟乙烯滤膜过滤, 转移滤饼至索氏萃取器中继续用丙酮萃取 10 天后将萃取液与滤液合并,所得溶液含 AEF1 和少量 NMP。以同样的方法用丙酮直接萃取煤样,得到含 AEF2 的溶液。用 GC/MS 分别分析上述溶液。在 80 ℃下真空干燥 MSEF 和煤样的丙酮萃余物 24 h 并在保持真空的状态下冷却至室温后取出称重,根据上式换算得出 AEF1 和 AEF2 的收率。

2 结果与讨论

2.1 MSEF、AEF1 和 AEF2 收率的比较

如图 1 所示,由各煤样所得 MSEF、AEF1 和 AEF2 收率大小的排列顺序分别为:FCC(33.5%)＞SFC(30.8%)＞HDGC(20.4%)、SFC(8.3%)＞HDGC(5.2%)＞FCC(4.1%) 和 HDGC(5.1%)＞SFC(4.8%)＞FCC(2.9%)。如表 1 所示,3 种煤样的 C 和 H 含量很接近,H/C 原子比也相差不大,但所得各萃取物的收率有明显差别。尽管由 FCC 所得 MSEF 的收率最高,但所得 AEF1 和 AEF2 的收率却最低,应该由 3 种煤组成结构的差别所致。由相同煤样所得 AEF1 的收率大于 AEF2 的收率,可能是在直接用丙酮萃取煤样时,煤样中部分丙酮可溶的成分被包裹在丙酮不溶的成分中,形成"胶囊效应",而 CS_2/NMP 混合溶剂能够溶解该"胶囊",从而释放出其中的丙酮可溶成分。

2.2 AEF1 和 AEF2 的 GC/MS 分析

用 GC/MS 分析了由各煤样所得 AEF1,意外地从中检测出一系列质谱基峰或次峰 m/z 为 98 的成分。图 2 至图 4 分别给出由 SFC、HDGC 和 FCC 所得 AEF1 的总离子流谱图(total ion chromatogram,简称 TIC)和提取 m/z 98 的选择离子流色谱图(selective ion chromatogram,简称 SIC),相应的质谱图分别示于图 5 至图 7。从 SFC 所得 AEF1 中检测出的 11 种成分(图 5 中的 SFC-AEF1-1 至 SFC-AEF1-11)、从 HDGC 所得 AEF1 中检测出的 5 种成分(图 6 中的 HDGC-AEF1-1 至 HDGC-AEF1-5)和从 FCC 所得 AEF1 中检测出的 4 种成分(图 7 中的 FCC-AEF1-1、FCC-AEF1-2、FCC-AEF1-4 和 FCC-AEF1-5)中 m/z 为 98 的峰均为基峰,即使 m/z 为 98 的次峰(图 7 中的 FCC-AEF1-3 和 FCC-AEF1-6)的相对丰度也大于 75%。然而,在所用的 Nist 和 Wiley 谱库中没有检索到与这些成分相匹配的任何质谱,且在本课题组已经研究过的数十种国内外煤样的萃取物中从未检测到这些成分。如图 8 所示,NMP 本身的质谱基峰和次峰的 m/z 分别是 99 和 44,因而所检测的成分肯定不是 NMP。

分别以 $A_{SIC\,(max)}$ 和 $A_{TIC\,(max)}$ 表示 SIC 最高峰和对应的 TIC 峰的丰度,由 SFC、HDGC 和 FCC 所得 AEF1(图 2 至图 4)的 $A_{SIC\,(max)}/A_{TIC\,(max)}$ 值分别为 37.5%、26.0% 和 62.4%,而由 SFC、HDGC 和 FCC 所得 AEF2(图 9 至图 11)的 $A_{SIC\,(max)}/A_{TIC\,(max)}$ 值分别为 0.9%、2.3% 和 1.8%,且图 9 至 11 所示的 SIC 各峰的质谱中 m/z 98 峰的最高相对丰度小于 10%(图 12 中给出代表性的示例)。这些事实说明 AEF1 与 AEF2 的组成差别很大,AEF1 中质谱基峰或次峰 m/z 为 98 的成分不是这些煤样中固有的成分,而应该与所用的 NMP 有关。

2.3 AEF1 中质谱基峰或次峰 m/z 为 98 的成分形成机理的分析

考察图 5 至图 7 可知,AEF1 中质谱基峰或次峰 m/z 为 98 的成分的分子离子 m/z 均远大于 98,也就是说,m/z 为 98 的质谱基峰或次峰是这些成分的碎片离子峰。由于 NMP 的分子量为 98 且 NMP 分子中含有一H,所检测到的质谱基峰或次峰 m/z 为 98 的碎片离子应该是 NMP

失去—H 形成的。但是,如图 8 所示,NMP 的质谱基峰和次峰的 m/z 都不是 98,说明在 GC/MS 分析过程中 NMP 本身失去—H 并不十分容易,所检测的 AEF1 中质谱基峰或次峰 m/z 为 98 的成分只能是 NMP 与煤中某些 OS 缔合的产物,即 NMP-OS 缔合体。

如图 13 所示,若 NMP 与 OS 间的缔合作用不强,在 GC/MS 分析过程中 NMP-OS 缔合体在进入 GC 气化室时就可能立刻受热解离;若 NMP 与 OS 间的缔合作用很强,在 GC/MS 分析过程中 NMP-OS 缔合体有可能通过 GC 的毛细管色谱柱进入质谱检测器被离子化,形成 M_1^+ 或 M_2^+,且只有形成 M_1^+ 时才能检测到 m/z 98 为高丰度(基峰或次峰)的碎片离子。换言之,从 3 种国产烟煤所得 AEF1 中检测出一系列质谱基峰或次峰 m/z 为 98 的成分的实验事实提供了 NMP 与这些烟煤中某些 OS 强烈缔合的证据。

3 结论

通过 GC/MS 分析在 3 种国产烟煤所得 AEF1 中检测出一系列质谱基峰或次峰 m/z 为 98 的成分,而在 AEF2 中没有检测出这些成分,且 NMP 的质谱基峰和次峰的 m/z 都不是 98,这些事实说明 NMP 与这些烟煤中某些 OS 发生了强烈的缔合作用。

Mg_xMn_y/ZrO_2 催化剂上乙腈和甲醇选择性合成丙烯腈的研究

摘要

用浸渍法制备了一系列不同 Mn 负载量的 $MgMn/ZrO_2$ 催化剂。这些催化剂用作由乙腈与甲醇反应合成丙烯腈。适量 Mn 的添加能够提高催化剂的活性,也能够提高丙烯腈的选择性。经检测,Mn 最适宜的量为 Mn/Zr = 0.2(摩尔比),在该催化剂上丙烯腈的产率能够达到 13.6%。通过各种表征,例如:XRD,XPS,SEM,CO_2—TPD 和 N_2 物理吸附实验,表明 Mn 掺杂后能够改善催化剂性质,这主要是由于 Mg_2MnO_4 的生成和催化剂上强碱性位的存在。然而,当 Mn 的含量高时,催化剂的催化活性降低,这主要是由于催化剂的活性中心被非活性物质 Mn_2O_3 所覆盖以及催化剂碱性中心相对较弱所致。

关键词:$MgMn/ZrO_2$ 催化剂;丙烯腈;乙腈;甲醇

1 引言

丙烯腈是一种非常重要的化工原料,在聚合工业界更为重要。通常,它是在多组分铋、钼催化剂上由丙烯氨氧化制得。乙腈是丙烯氨氧化制丙烯腈的主要副产物,尽管乙腈能够经过典型的腈化反应生成胺类、氨基化合物、腈的卤化物、酮类以及其他物质,但可用量极小。目前,乙腈除主要用作溶剂外,大量过剩的乙腈被燃烧处理。因此,利用乙腈合成更为有用的化学物品是一种非常重要的技术。

最近几年,几个研究小组试图通过在 CH_3CN 的 α 位上引入碳原子,以将 CH_3CN 转化为更有价值的丙烯腈。甲烷、甲醇和甲醛可用作反应的甲基化试剂,稳定性较好的甲烷使得该反应需在较高温度下进行,甲醇由于它的高活性而常用作甲基化试剂,同时甲醇还可以由天然气制得,其供应充足。既然乙腈分子有吸电子基团(—CN),这就要求用碱催化剂来引发乙腈的α碳原子,从而产生我们想要的甲基化反应。碱金属负载在硅胶上,碱金属或碱土金属负载在二氧化硅上以及过渡金属元素负载在氧化镁上,都可以用作乙腈与甲醇反应合成丙烯腈的催化剂。上述催化剂催化活性顺序为 Li > Na > K > Rb,但是丙烯腈产率仍旧很低。MgO 负载 Cr,Fe,Mn 和 Cu 等过渡金属后,在 350 ℃ 到 400 ℃ 范围内可以将乙腈与甲醇选择性的转换成丙烯腈。然而,据报道乙腈的最大转化率在 10% 左右。因此,为了提高该反应的转化率以及选择性,有必要寻找新的催化剂。

在这篇文章中,我们制备了一组由 Mn 改性的 MgO 负载在 ZrO_2 上用来作为由乙腈和甲醇反应合成丙烯腈的催化剂。可以看到这些催化剂显示出很好的活性以及选择性。根据对这些催化剂结构以及 Mn 含量对催化结果的影响,讨论了催化剂的性质。

2 实验部分

2.1 催化剂的制备

ZrO_2 负载 Mg 和 Mn 元素的催化剂用浸渍法制备,将 ZrO_2 浸渍在 $Mg(NO_3)_2$ 和 $Mn(NO_3)_2$ 的混合溶液中 2 h 后,将该悬浮液保持在 95 ℃ 并不断搅拌直至水分完全蒸发,将获得的糊状物放于烘箱中在 120 ℃ 烘干 4 h,随后在 600 ℃ 下焙烧 5 h。将制得的催化剂记作 Mg_xMn_y/ZrO_2(x 和 y 分别代表 Mg/Zr 和 Mn/Zr 的摩尔比)。

2.2 催化性能实验

乙腈和甲醇反应合成丙烯腈在空压下,于气相反应中进行,用石英管($\phi 8\ mm \times 215\ mm$)作为反应器。对于每一催化反应在石英管中间装填大约 500 mg 催化剂,催化剂两边用石英棉固定。在 70 mL/min 的 N_2 气流保护下将催化剂加热至反应温度,反应物($CH_3CN : CH_3OH$ 摩尔比为 1:10 的反应混合液)经由一高压恒流泵注入到预热管线中,反应混合液流量为 0.03 mL/min(再由流量为 70 mL/min 的氮气稀释后带入反应器中反应)。反应产物采用配有 PEG 20M 毛细管色谱柱的 7890 II 型气相色谱仪(FID)分析。目的产物丙烯腈(用 AN 表示),丙腈(用 PN 表示)和甲基丙烯腈(用 MAN 表示)。图 5 到图 8 所提及乙腈转化率和丙烯腈选择性通过含氮产物的归一法来计算,即通过下面公式计算所得:

$$转化率 = (([AN] + [PN] + [MAN]) / [Acetonitrile]_{reactant}) \times 100\% \quad (1)$$

$$选择性 = ([AN] / ([AN] + [PN] + [MAN])) \times 100\% \quad (2)$$

2.3 催化剂表征

表面积测定采用 BET 方法,在 -196 ℃,N_2 吸附,Micromeritics TriStar 3000 型物理吸附仪进行。催化剂的 XRD 谱峰在荷兰 Philips 公司的 PW3040/60 X 射线衍射仪上获得,用 Cu K_α 辐射(0.15418 nm),管电压和管电流分别是 40 kV 和 40 mA。催化剂的表面形貌由 SEM(Philips 公司 XL 30 型)获得。XPS 实验在美国 PHI 公司的 PHI 5000C 上进行;采用条件为铝靶(1486.6 eV)。所有的结合能参照 C1s = 284.6 eV 为基准进行结合能校正,其误差不超过 ±0.2 eV。

CO_2 脱附峰由下列方法获得,试样先经 500 ℃ 通 He 预处理 30 min,冷却至室温吹扫 0.5 h 后,CO_2 脉冲进样吸附饱和至峰面积保持不变。(He 再吹扫 1 h 后进行 TPD 实验),最高脱附温度为900 ℃,升温速率20 ℃/min,脱附物质用质谱仪(Balzers公司的Omnistar 200型)进行分析。

3 结果和讨论

3.1 $Mg_{0.5}Mn_y/ZrO_2$催化剂的比表面和形貌分析

表1所示为BET方法测定的ZrO_2载体及$Mg_{0.5}Mn_y/ZrO_2$催化剂的比表面积、孔体积及平均孔径大小。从表中可以看出,ZrO_2载体和$Mg_{0.5}Mn_y/ZrO_2$催化剂的比表面积相对较低,这是由载体性质及高温(600 ℃)焙烧所共同造成的。ZrO_2载体的比表面积为16.1 m^2/g,孔容为0.1463 cm^3/g,孔径平均大小为36.3 nm,当Mg负载在ZrO_2载体上后,比表面积及孔体积大小均约为原来的1/4;平均孔径大小却在原来的基础上增加了4.6 nm。这可能是由于负载Mg组分后,发生部分堵孔现象,尤其是对孔径较小的孔堵塞较为明显。同时,从表中我们还可以看到,随着Mn组分负载量的增加,催化剂$Mg_{0.5}Mn_y/ZrO_2$的比表面积变大,孔径减小。或许是由于催化剂中对其比表面的贡献主要反映在微孔区。当PH=10时,这与Trunschke等用La_2O_3/ZrO_2浸渍$(NH_4)_2CrO_4$水溶液制备La—Cr/ZrO_2催化剂时所出现的结论基本一致。

图1给出了ZrO_2、$Mg_{0.5}/ZrO_2$、$Mg_{0.5}Mn_{0.2}/ZrO_2$和$Mg_{0.5}Mn_{0.4}/ZrO_2$催化剂样品的扫描电镜图像,这些催化剂是在600 ℃处理5.0 h所得到的。我们可以清楚地看到,当表面光洁的ZrO_2载体负载Mg物种后,在催化剂表面形成一薄层含有Mg的膜状物。由(b)图可知,尽管这些膜状物厚度相差不大(大约为40 nm),但其尺寸大小分布范围在0.4~2.1 μm之间。同时,从(c)图也可以看到,催化剂$Mg_{0.5}Mn_{0.2}/ZrO_2$表面也有一层厚度基本一致的膜状物。可是,这些膜状物大小基本一致,约为0.9 μm。(d)图显示:$Mg_{0.5}Mn_{0.4}/ZrO_2$催化剂的表面膜状物较厚。

3.2 $Mg_{0.5}Mn_y/ZrO_2$催化剂的结构表征

图2为$Mg_{0.5}Mn_y/ZrO_2$催化剂在600 ℃下焙烧后所得的XRD谱图。通过与标准卡片对比可知,在$Mg_{0.5}/ZrO_2$催化剂中ZrO_2呈单斜态,MgO成立方晶相。谱峰中并没有MgO和ZrO_2化合物的衍射峰。从由Mn改性的$Mg_{0.5}/ZrO_2$催化剂的XRD谱图中可以看到:出现了35.6°、43.3°和62.8°的衍射峰,它们均是Mg_2MnO_4衍射所产生的主要谱峰。通过比较35.6°和42.9°两个较强谱峰的强度,二者分别对应Mg_2MnO_4和MgO物相,我们可以得出结论:随着Mn负载量的增加,催化剂中Mg_2MnO_4与MgO的摩尔比也在增大。可以认为Mn含量的增大能够导致Mg_2MnO_4的生成,而该组分正是由乙腈与甲醇反应合成丙烯腈的活性物中。当催化剂中Mn/Mg比值大于1:2(在化合物Mg_2MnO_4中的原子比)时,例如在$Mg_{0.5}Mn_{0.3}/ZrO_2$和$Mg_{0.5}Mn_{0.4}/ZrO_2$催化剂中,XRD谱图中MgO的谱峰消失,而代之以Mn_2O_3的特征谱峰(32.9°),这表明,在这两个催化剂中除了有Mg_2MnO_4和ZrO_2之外,出现Mn_2O_3物相(图2e, 2f)。

Deraz等将碳酸镁浸入硝酸锰溶液中,并在400~800 ℃焙烧制备MgO负载活性物种Mn催化剂时,发现当焙烧温度升高到600 ℃时,MnO_2完全转化为Mn_2O_3和Mg_2MnO_4,其结果和我们的实验结果一致。通过研究,他认为部分Mn_2O_3进一步发生了如下反应:

$$4MgO + Mn_2O_3 + 1/2O_2 \longrightarrow 2Mg_2MnO_4$$

3.3 XPS表征结果

图3所示为6个催化剂的XPS谱图。从中可以看到所有样品均产生Zr 3d5/2和Zr 3d3/2特征谱峰,结合能分别为182.2 eV和184.6 eV。它表明Zr在$Mg_{0.5}Mn_y/ZrO_2$催化剂中均以

ZrO_2 状态存在。值得注意的是,很难观察到 Mg 2p 谱峰,这是由于 Zr 4s 和 Mn 3p 谱峰的干扰所造成。因此,我们仅以 Mg 2s 谱峰为 Mg 存在状态的研究依据。在所有催化剂中都出现 89.4 eV 的 Mg 2s 特征谱峰,证明 Mg 物种的主要存在形式为 Mg(Ⅱ)。该结果与 XRD 中 Mg 以 MgO 和 Mg_2MnO_4 两种状态存在相符合。从 Mn 2p 谱峰中,我们可以清楚地观察到谱峰强度随着 Mn 负载量的增加而加大。在 $Mg_{0.5}Mn_{0.05}/ZrO_2$,$Mg_{0.5}Mn_{0.1}/ZrO_2$ 和 $Mg_{0.5}Mn_{0.2}/ZrO_2$ 催化剂中 Mn 3d5/2 结合能为 642.1 eV,这表明 Mn 以 Mg_2MnO_4 状态存在;然而在 $Mg_{0.5}Mn_{0.3}/ZrO_2$ 和 $Mg_{0.5}Mn_{0.4}/ZrO_2$ 催化剂中 Mn 3d5/2 结合能向低处位移至 641.6 eV,表明有 Mn 的其他物种产生,通过对谱峰曲线进行拟合,我们认为 641.6 eV 和 642.1 eV 的谱峰分别被归属为 Mn_2O_3 和 Mg_2MnO_4 所产生。这个结论进一步证实了当 Mn/Mg 的值较小时,Mn 主要以 Mg_2MnO_4 状态存在,而当 Mn/Mg 值增大时,我们可以看到在 $Mg_{0.5}Mn_y/ZrO_2$ 系列催化剂中 Mn_2O_3 和 Mg_2MnO_4 两种化合物生成。

通过一个简单计算,我们可以知道,在 Mn_2O_3 和 Mg_2MnO_4 中,Mg/Mn 的值分别为 0 和 2。如果所有的 Mn 与 Mg 均以 Mn_2O_3 和 Mg_2MnO_4 两种状态存在,则在 $Mg_{0.5}Mn_{0.3}/ZrO_2$ 催化剂中 Mn_2O_3 与 Mg_2MnO_4 的比例为 1:5,在 $Mg_{0.5}Mn_{0.4}/ZrO_2$ 催化剂中二者的比例为 3:5。按照图 3 所示,在 $Mg_{0.5}Mn_{0.3}/ZrO_2$ 和 $Mg_{0.5}Mn_{0.4}/ZrO_2$ 催化剂中 Mn 3d 的两个谱峰面积之比分别为 0.22 和 0.61,这与计算结果吻合较好。

3.4 $Mg_{0.5}Mn_y/ZrO_2$ 催化剂的碱性

图 4 为 $Mg_{0.5}Mn_y/ZrO_2$ 催化剂的 CO_2—TPD 谱图。从图中可以看出,$Mg_{0.5}/ZrO_2$ 催化剂在 114 ℃ 和 646 ℃ 有两种强度不同的脱附峰,峰强度之比大约为 3:1,表明该催化剂有两种强度不同的碱中心。Jiang 等做 MgO/ZrO 催化剂的 CO_2—TPD 时也发现了该低温脱附峰,但他们并没有报道该催化剂的高温脱附峰。

对于由 Mn 改性的催化剂来说,我们也得到了相同峰强、相同脱附温度的低温脱附谱峰。另一方面,当添加了少量的 Mn 组分后,催化剂的高温脱附峰向高温方向位移,且峰强度变大,(催化剂 $Mg_{0.5}Mn_{0.2}/ZrO_2$ 的高温脱附峰达到最大)此后,随着 Mn 含量的增加,峰强降低,峰向低温方向偏移。这表明,对于 $Mg_{0.5}Mn_y/ZrO_2$ 催化剂来说,在一定范围内,随着 Mn 的添加,催化剂的强碱性位增加。

3.5 催化剂的催化性能表征

由乙腈与甲醇反应合成丙烯腈是一个复杂的反应,反应过程中甲醇容易分解,尽管少量的甲醇转化成了甲烷和一氧化碳,这一点在 $Mg_{0.5}Mn_y/ZrO_2$ 作为催化剂的反应中可以忽略,主要是由于甲醇与乙腈的摩尔比为 10:1,甲醇大大过量。反应产物通过气相色谱和质谱仪分析,主要包括丙烯腈,丙腈以及甲基丙烯腈。这样,该反应的转化率和选择性通过以上方程(1)和(2)计算出来既是合理的。

3.5.1 Mg 负载量对催化剂性能的影响

图 5 给出了反应温度 480 ℃ 时,在 $Mg_xMn_{0.2}/ZrO_2$ 催化剂存在下,对于乙腈与甲醇反应,Mg 负载量对催化剂性能的影响。所研究的催化剂为 $Mg_{0.4}Mn_{0.2}/ZrO_2$,$Mg_{0.5}Mn_{0.2}/ZrO_2$,$Mg_{0.6}Mn_{0.2}/ZrO_2$,$Mg_{0.7}Mn_{0.2}/ZrO_2$ 和 $Mg_{0.8}Mn_{0.2}/ZrO_2$。为了揭示 Mg 负载量的影响,所有反应条件固定为:装填 500 mg 催化剂,N_2 的流速为 70 mL/min,反应物(乙腈 + 甲醇)流速为 0.03 mL/min,甲醇/乙腈的摩尔比为 10:1,反应温度为 480 ℃。在这些条件下,转化率与选择性变化都不大(分别约为 16.2% 和 84.0%)。

3.5.2 反应温度对催化剂性能的影响

图 6 为温度对 $Mg_{0.5}Mn_{0.2}/ZrO_2$ 催化剂上乙腈转化率及丙烯腈选择性的影响结果。对于乙腈的转化率和丙烯腈的选择性有不同的最佳温度。乙腈转化率的最适宜温度为 480 ℃，而丙烯腈选择性的最佳温度为 400 ℃。综合考虑乙腈的转化率以及丙烯腈的选择性这两个方面，480 ℃ 的反应温度是相对较好的温度，在此温度下丙烯腈的产率为 13.6%。

3.5.3 反应时间对催化剂性能的影响

对于乙腈与甲醇反应的稳定性也作了检测。图 7 所示为反应温度 480 ℃，在催化剂 $ZrMg_{0.5}Mn_{0.2}$ 上，反应时间对乙腈转化率及丙烯腈选择性影响的结果。从图 7 可以看出，反应开始时，乙腈的转化率降低，而丙烯腈的选择性则略微增加，3 h 后基本达到稳定状态。当稳定性测试进行到 8 h 后，可以发现催化剂活性已经非常低。活性降低的主要原因看上去是催化剂表面积碳造成的，因为催化剂的颜色由最初的灰白变为暗黑色。为了证实这一点，我们将反应 8 h 后的催化剂进行氧化激活：将失活的催化剂在 600 K 温度下煅烧 2 h，紧接着在 N_2 保护下，温度保持不变时活化 2 h。再生催化剂显示出与新鲜催化剂相同的活性，(从而证实了催化剂失活是表面积炭所造成的这一结论)。

3.5.4 Mn 负载量对催化剂性能的影响

图 8 给出了在 480 ℃ 时乙腈转化率与丙烯腈选择性随催化剂 $Mg_{0.5}Mn_y/ZrO_2$ 中 Mn 含量变化的关系。从图 8 可以看出，催化剂 $Mg_{0.5}/ZrO_2$ 实际上是没有活性的，乙腈的转化率仅为 1% 左右。据报道：在 350 ℃ 和 400 ℃ 下，在没有负载任何其他物质的 MgO 做催化剂时由乙腈与甲醇反应合成丙烯腈时也得到了相似的结论。Mn 负载后催化剂的活性从图 8 可以看到，随着 Mn 含量的增加，乙腈转化率首先升高，继而减小。对于 $Mg_{0.5}Mn_{0.2}/ZrO_2$ 催化剂，转化率最大，约为 16.4%，其值大于 Mn—MgO 催化剂的转化率(9.6%)。随着 Mn 含量的继续增加，转化率略有降低。同时，我们从图 8 也可以看出，Mn 的最佳含量为 Mn/Zr = 0.2，(摩尔比)。此时该催化剂对丙烯腈的选择性为 84.0%，高于 $Mg_{0.5}/ZrO_2$ 催化剂的选择性(为 79.8%)，也高于 $Mg_{0.5}Mn_{0.4}/ZrO_2$ 催化剂的选择性(为 79.7%)。

从图 8 我们也可以看出：适量 Mn 的添加能够改善 $Mg_{0.5}Mn_y/ZrO_2$ 催化剂的活性。首先，从以上表征我们可以知道，$Mg_{0.5}Mn_{0.2}/ZrO_2$ 催化剂的高活性主要归因于 Mg_2MnO_4 物种的生成。按照 XPS 及 XRD 表征结果，Mn 的加入促使 MgO 生成 Mg_2MnO_4，相对于 MgO 表面，反应物分子较易吸附在 Mg_2MnO_4 催化剂表面上。然而，随着催化剂中 Mn 含量的增加，尤其是当 Mn/Mg 的摩尔比大于 1∶2 时，会导致非活性物种 Mn_2O_3 生成，Mn_2O_3 部分覆盖在活性物种之上，导致在催化剂上由乙腈与甲醇反应合成丙烯腈的活性降低。这一点从表 1 我们也可以看出，$Mg_{0.5}Mn_y/ZrO_2$ 催化剂的表面积随着 Mn 含量的增加而增加。其次，从 $Mg_{0.5}Mn_y/ZrO_2$ 催化剂的 CO_2—TPD 谱图我们也可以看出，$Mg_{0.5}Mn_{0.2}/ZrO_2$ 催化剂的高温脱附峰的强度是最大的，对应着它有一个最强的碱性中心，强碱性对于乙腈与甲醇脱氢来说，能够增加质子的提取能力，从而导致反应的两个中间态较为容易生成。最后，大的比表面积有利于反应物与催化剂活性部位的碰撞，提高乙腈的转化率。

4 结论

研究了一系列用浸渍法制备的，由 Mn 改性 $MgMn/ZrO_2$ 催化剂的物理化学性质表明，当 Mn 添加到催化剂中能够提高催化剂的表面积以及增加催化剂的碱性中心，同时也能够促使 Mg_2MnO_4 和 Mn_2O_3 的生成。对于由乙腈与甲醇反应合成丙烯腈这个反应，$MgMn/ZrO_2$ 催化剂是一个有效的催化剂。当催化剂中添加适当 Mn 组分后,生成的 Mg_2MnO_4 含量较高,催化剂表面的强碱中心数较多时,有利于乙腈与甲醇选择性合成丙烯腈。而当 Mn 的最适宜的含量为 Mn/Zr = 0.2(原子个数比),Mg/Zr 原子比例是 1∶2 时,催化剂 $Mg_{0.5}Mn_y/ZrO_2$ 的活性最佳,在 480 ℃时丙烯腈的选择性为 84.0%,乙腈的转化率为16.2%。

Names of the Chemical Elements

化学元素名称

第一主族元素
- H 氢 Hydrogen
- Li 锂 Lithium
- Na 钠 Sodium
- K 钾 Potassium
- Rb 铷 Rubidium
- Cs 铯 Cesium
- Fr 钫 Francium

第二主族元素
- Be 铍 Beryllium
- Mg 镁 Magnesium
- Ca 钙 Calcium
- Sr 锶 Strontium
- Ba 钡 Barium
- Ra 镭 Radium

第三主族元素
- B 硼 Boron
- Al 铝 Aluminum
- Ga 镓 Gallium
- In 铟 Indium
- Tl 铊 Thallium

第四主族元素
- C 碳 Carbon
- Si 硅 Silicon
- Ge 锗 Germanium
- Sn 锡 Tin
- Pb 铅 Lead

第五主族元素
- N 氮 Nitrogen
- P 磷 Phosphorus
- As 砷 Arsenic
- Sb 锑 Antimony
- Bi 铋 Bismuth

第六主族元素
- O 氧 Oxygen
- S 硫 Sulfur
- Se 硒 Selenium
- Te 碲 Tellurium
- Po 钋 Polonium

第七主族元素
- F 氟 Fluorine
- Cl 氯 Chlorine
- Br 溴 Bromine
- I 碘 Iodine
- At 砹 Astatine

第零族元素
- He 氦 Helium
- Ne 氖 Neon
- Ar 氩 Argon
- Kr 氪 Krypton
- Xe 氙 Xenon
- Rn 氡 Radon

第一过渡系元素
- Sc 钪 Scandium
- Ti 钛 Titanium
- V 钒 Vanadium
- Cr 铬 Chromium
- Mn 锰 Manganese
- Fe 铁 Iron
- Co 钴 Cobalt
- Ni 镍 Nickel
- Cu 铜 Copper
- Zn 锌 Zinc

第二过渡系元素
- Y 钇 Yttrium
- Zr 锆 Zirconium
- Nb 铌 Niobium
- Mo 钼 Molybdenum
- Tc 锝 Technetium
- Ru 钌 Ruthenium
- Rh 铑 Rhodium
- Pd 钯 Palladium
- Ag 银 Silver

Cd 镉 Cadmium

第三过渡系元素

La 镧 Lanthanum	Hf 铪 Hafnium	Ta 钽 Tantalum
W 钨 Tungsten, (Wolfram)		Re 铼 Rhenium
Os 锇 Osmium	Ir 铱 Iridium	Pt 铂 Platinum
Au 金 Gold	Hg 汞 Mercury	

镧系元素

La 镧 Lanthanum	Ce 铈 Cerium	Pr 镨 Praseodymium
Nd 钕 Neodymium	Pm 钷 Promethium	Sm 钐 Samarium
Eu 铕 Europium	Gd 钆 Gadolinium	Tb 铽 Terbium
Dy 镝 Dysprosium	Ho 钬 Holmium	Er 铒 Erbium
Tm 铥 Thulium	Yb 镱 Ytterbium	Lu 镥 Lutetium

锕系元素

Ac 锕 Actinium	Th 钍 Thorium	Pa 镤 Protactinium
U 铀 Uranium	Np 镎 Neptunium	Pu 钚 Plutonium
Am 镅 Americium	Cm 锔 Curium	Bk 锫 Berkelium
Cf 锎 Californium		

Names of the Frequently-used Experimental Instruments

常用化学实验仪器名称

中文	English	中文	English
普通试管	test tube	试剂瓶	reagent bottle
离心试管	centrifuge tube	普通圆底烧瓶	round flask
试管架	test-tube rack	磨口圆底烧瓶	ground-in round flask
试管夹	test-tube clamp	蒸馏烧瓶	distilling flask
泥三角	pipeclay triangle	分液漏斗	separating funnel
温度计	thermometer	滴液漏斗	dropping funnel
烧杯	beaker	冷凝管	condenser
毛刷	hair brush	蒸馏头	distilling head
锥形瓶	erlenmeyer flask	塞子	stopper
量筒	measuring cylinder	应接管	distillation adapter
漏斗	funnel	坩埚	crucible
吸量管、移液管	pipette	蒸发皿	evaporating dish
容量瓶	volumetric flask	研钵	mortar
酸式滴定管	acid burette	药匙	medicine spoon
碱式滴定管	alkali burette	点滴板	spot plate
吸滤瓶	filter flask	水浴锅	water bath kettle
布式漏斗	Buchner funnel	三脚架	tripod
称量瓶	weighing bottle	石棉网	asbestos center gauze
表面皿	watch glass	铁架台	iron stand
滴定管	burette（buret）	坩埚钳	crucible tongs
T形管	T joint	碘量瓶	iodine flask
Y形接头	Y joint	燃烧匙	combustion spoon
U形管	U-tube	砂芯漏斗	glass sand funnel / sand core funnel / core print funnel
三口烧瓶	three-neck flask	集气瓶	gas-jar
冷阱	cold trap		
干燥管	drying tube		
干燥器	desiccator		

References
参 考 文 献

第一部分

[1] 武汉大学,吉林大学. 无机化学[M]. 北京:高等教育出版社,1994.

[2] Jorissen W P, Bassett H, Damiens A, et al. 国际无机化合物命名法则的中译[J]. 化学通报,1953,3:111-123.

[3] 丁欣宇,华平,景晓辉. 浅析化工专业英语特点及教学方法[J].南通航运职业技术学院学报,2007,6(2):103-105.

[4] 那顺孟和,杨秋林. 浅析无机二元化合物的中英文命名法的关系[J].内蒙古石油化工,2007,8:320-321.

[5] 余丽萍,汪萍,杨斌,等. 双语教学中无机化合物命名的教学探索[J]. 北京大学学报:哲学社会科学版,2007,5:169-171.

[6] 邢其毅,裴伟伟,徐瑞秋,等. 基础有机化学[M]. 北京:高等教育出版社,2005.

[7] 邵万明. 美国化学文摘的有机化合物命名原则[J].图书情报工作,1987,4:40-44.

[8] 唐步芳. 有机化合物系统命名法的教学[J]. 成都大学学报:自然科学版,1987,1:66-68.

[9] 王述楠. 谈常见有机化合物系统命名规律的简化[J].江西教育学院学报,1994,15(6):72-80.

[10] 杨定乔,龙玉华,王升富. 化学化工专业英语[M]. 北京:化学工业出版社,2008.

第二部分

[1] Clayden J. Organic Chemistry [M]. USA: Oxford University Press, 2000.

[2] Vollhardt K, Peter C. Organic Chemistry: Structure and Function [M]. 4th ed. New York: W. H. Freeman & Co Ltd, 2002.

[3] 大连理工大学分析化学教研室. 分析化学[M]. 大连:大连理工大学出版社,2008.

[4] Skoog D A. Analytical chemistry: an introduction [M]. 7th ed. Fort Worth: Saunders College Pub., 2000.

[5] Hong T K, Koo B H, Ly S Y, Kim M H, Czae M Z. Use of pH electrode for precipitation titration analysis: Theory and practice [J]. Journal of Analytical Chemistry, 2009, 64(11): 1158-1165.

[6] Esteban M, Arino C, Dlaz-cruz J M. Chemometrics in electroanalytical chemistry [J]. Critical Reviews in Analytical Chemistry, 2006, 36(3-4): 295-313.

[7] Straughan B P, Walker S. Spectroscopy [M]. 2nd ed. London: Chapman and Hall:

New York: Wiley, 1976.
- [8] Laidler K J, Meiser J H, Sanctuary B C. Physical chemistry [M]. 4th ed. Boston: Houghton Mifflin Co., 2003.
- [9] Dasguptas. Macromlecular structural study on radiation induced terpolaymers of nylon [J]. Applied Spectroscopy, 1976, 21(6): 387-389.
- [10] Yan S D, Stern D M. Mitochondrial dysfunction and Alzheimer's disease: role of amyloid-beta peptide alcohol dehydrogenase (ABAD) [J]. International Journal of Experimental Pathology, 2005, 86(3): 161-171.
- [11] Shahinpoor M. Microelectromechanics of ionic polymeric gels as electrically controllable artificial muscles [J]. Journal of Intelligent Material Systems and Structures, 1995, 6(3): 307-314.
- [12] Chen I J, Taneja R, Yin D X. Chemical substituent effect on pyridine permeability and mechanistic insight from computational molecular descriptors [J]. Molecular Pharmaceutics, 2006, 3(6): 745-755.
- [13] Mandal A, Parkash M, Kumar R M. Ab Initio and DFT studies on methanol-water clusters [J]. Journal of Physical Chemistry, 2010, 114(6): 2250-2258.
- [14] Fitzgerald G, Andzelm J. Chemical applications of density functional theory comparison to experiment, Hartree-Fock, and perturbation theory [J]. Journal of Physical Chemistry, 1991, 95(26): 10531-10534.
- [15] Young D C. Computational Chemistry: A Practical Guide for Applying Techniques to Real World Problems [M]. New York: John Wiley & Sons Inc., 2002.
- [16] Guy H, Grant W, Graham R. Computational chemistry [M]. New York: Oxford University Press, 1995.
- [17] Speight J G. Lange's handbook of chemistry [M]. 16th ed. New York: McGraw-Hill Co., 2005.
- [18] 董坚. 化学化工专业英语[M]. 杭州:浙江大学出版社, 2010.
- [19] 张裕平, 姚树文, 龚文君. 化学化工专业英语[M]. 北京:化学工业出版社, 2007.
- [20] Yale Department of Chemistry. Chemistry Department Safety Manual [EB/OL]. [2008-09-01] http://chem.yale.edu/res/safety.html.
- [21] 姚日生,董岸杰,刘永琼.药用高分子材料[M].北京:化学工业出版社,2008.
- [22] 杨华明,宋晓岚,金盛明.新型无机材料[M].北京:化学工业出版社,2005.
- [23] 陈忠红,刘佳,陈琼,等.高分子压电复合材料研究进展[J].化工新型材料,2016,44:19-21.
- [24] 任学晖,李鑫,张田.浅析高分子化工材料在我国的发展[J].新材料与新技术,2016,42:38.
- [25] 江东亮.透明陶瓷——无机材料研究与发展重要方向之一[J].无机材料学报,2009,24:873-881.
- [26] 高金良,袁泽明,尚宏伟,雍辉,祁焱.氢储存技术及其储能应用研究进展[J].金属功能材料,2016,23:1-11.

第三部分

- [1] Whitesides G M. Whitesides group: Writing a Paper [J]. Advanced Materials, 2004,

参考文献

16(15): 1375-1377.

[2] Zhao L H, Zhao J J, Luo M F, Liu G Y, Song J, Zhang Y P. Acrylonitrile synthesis from acetonitrile and methanol over MgMn/ZrO$_2$ catalysts [J]. Catalysis Communications, 2005, 6: 617-623.

[3] Luo M F, Song Y P, Lu J Q, Wang X Y, Pu Z Y. Identification of CuO species in high surface area CuO - CeO$_2$ catalysts and their catalytic activities for CO oxidation [J]. Journal of Physical Chemistry C, 2007, 111(34): 12686-12692.

[4] Liu C M, Zong Z M, Jia J X, Liu G F, Wei X Y. An evidence for the strong association of N-methyl-2-pyrrolidinone with some organic species in three Chinese bituminous coals [J]. Chinese Science Bulletin, 2008, 53(8): 1157-1164.